Genes, Cells
A·N·D
Organisms

GREAT BOOKS IN
EXPERIMENTAL BIOLOGY

EDITED BY
John A. Moore

A Garland Series

17 TITLES THAT STAND AS
MONUMENTS OF BIOLOGICAL THOUGHT

GREAT BOOKS IN EXPERIMENTAL BIOLOGY

1. *THE CELL IN DEVELOPMENT AND HEREDITY*, by Edmund B. Wilson. 1925.

2. *THE CELL THEORY: A RESTATEMENT, HISTORY, AND CRITIQUE*, by John R. Baker. 1948–1955.

3. *SPECIES AND VARIETIES, THEIR ORIGIN BY MUTATION: LECTURES DELIVERED AT THE UNIVERSITY OF CALIFORNIA*, by Hugo De Vries. 3rd edition, 1912.

4. *THE GENETICS OF DROSOPHILA*, by Thomas Hunt Morgan (with Calvin B. Bridges and A. H. Sturtevant). 1925.

5. *THE THEORY OF THE GENE*, by Thomas Hunt Morgan. 1926.

6. *AN INTRODUCTION TO GENETICS*, by A. H. Sturtevant and G. W. Beadle. 1939.

7. *RECENT ADVANCES IN CYTOLOGY*, by C. D. Darlington. 1937.

8. *PHYSIOLOGICAL GENETICS*, by Richard Goldschmidt. 1938.

9. *INVERTEBRATE EMBRYOLOGY*, edited by Matazo Kume and Katsuma Dan. 1968.

10. *EMBRYONIC DEVELOPMENT AND INDUCTION*, by Hans Spemann. 1938.

11. *THE ORGANIZATION OF THE DEVELOPING EMBRYO*, by Ross G. Harrison. 1969.

12. *PROBLEMS OF FERTILIZATION*, by Frank R. Lillie. 1919.

13. *THE BIOLOGY OF THE CELL SURFACE*, by Ernest Just. 1939.

14. *FEATURES IN THE ARCHITECTURE OF PHYSIOLOGICAL FUNCTION*, by Joseph Barcroft. 1934.

15. *THE NEUROPHYSIOLOGICAL BASIS OF MIND; THE PRINCIPLES OF NEUROPHYSIOLOGY*, by John Carew Eccles. 1953.

16. *RECOLLECTIONS OF MY LIFE*, by Santiago Ramon Y. Cajal. 1937.

17. *THE DISCOVERY AND CHARACTERIZATION OF TRANSPOSABLE ELEMENTS: THE COLLECTED PAPERS OF BARBARA McCLINTOCK*. 1987.

THE GENETICS OF DROSOPHILA

Thomas Hunt Morgan

with

Calvin B. Bridges

and

A. H. Sturtevant

Garland Publishing, Inc.

NEW YORK & LONDON • 1988

Library of Congress Cataloging-in-Publication Data

Morgan, Thomas Hunt, 1866–1945.
The genetics of Drosophila.

(Genes, cells, and organisms)
Reprint. Originally published: 's-Gravenhage :
M. Nijhoff, 1925.
Bibliography: p.
Includes index.
1. Drosophila—Genetics. I. Bridges, Calvin B.
(Calvin Blackman), 1889–1938. II. Sturtevant, A. H.
(Alfred Henry), 1891– . III. Title. IV. Series.
[DNLM: 1. Drosophila—genetics. QX 505 M849g 1925a]
QL 537.D76M67 1988 595.77'4 88-24279
ISBN 0-8240-1383-2

All volumes in this series are printed on
acid-free, 250-year-life paper.

PRINTED IN THE UNITED STATES OF AMERICA

THE GENETICS OF DROSOPHILA

by

T. H. MORGAN, C. B. BRIDGES, A. H. STURTEVANT

With 62 figures and 6 Plates

TABLE OF CONTENTS

CHAPTER I

INTRODUCTION

Drosophila melanogaster (formerly D. ampelophila) had been used in various biological experiments beginning in 1905 with CARPENTER's study of its reaction to light, to gravity, and to agitation, before its value for genetic work was appreciated. The year following, 1906, an extensive study of the effect of inbreeding was published by CASTLE, CARPENTER, CLARK, MAST and BARROWS, showing that if the offspring of fertile pairs were selected in each generation, no decrease in fertility appeared as a result of close inbreeding. LUTZ reported in 1907 at the Zoölogical Congress, observations on the inheritance of a wing-

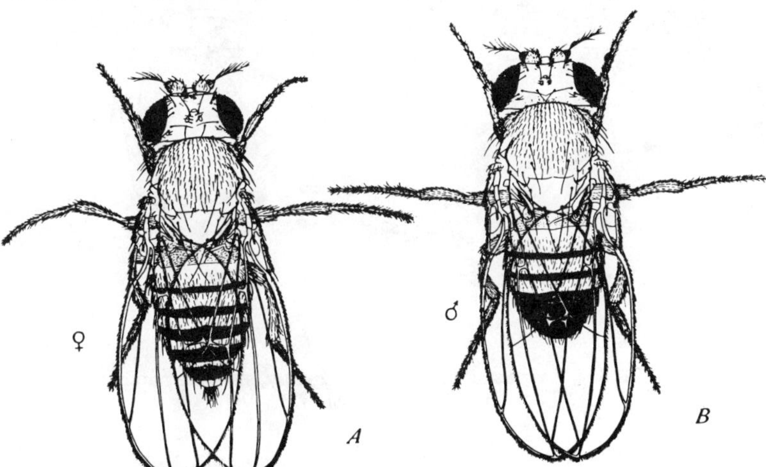

Fig. 1. The wild-type female (A) and male (B) of Drosophila melanogaster.

vein modification of Drosophila melanogaster, but his paper did not appear until 1911. DELCOURT ('09) reported observations on the inheritance of variations in the wing-veins in Drosophila confusa and described some "anomalies" in Drosophila ampelophila also. The results were "incoherent". Later, DELCOURT and GUYÉNOT ('10, '11), as well as GUYÉNOT ('17), in a separate paper, published an extensive study on the effects of food conditions, more especially the rôle of micro-organisms in the diet. This led finally to the elaboration of a method of rearing the flies under aseptic conditions — a method that later, in the hands of LOEB and NORTHROP, led to significant results [1]. MOENKHAUS ('11) also worked on the effects of inbreeding on fertility of Drosophila ampelophila and studied variations of the sex-ratio in relation to selection. An extended criticism of this paper, as well as that of CASTLE and his colaborators, is given by DELCOURT and GUYÉNOT, but since they have themselves added very little constructive work of their own in these fields and since the later work on Drosophila has cleared up some of the earlier differences, in directions not contemplated or approved of by DELCOURT and GUYÉNOT, their criticism need not be repeated here.

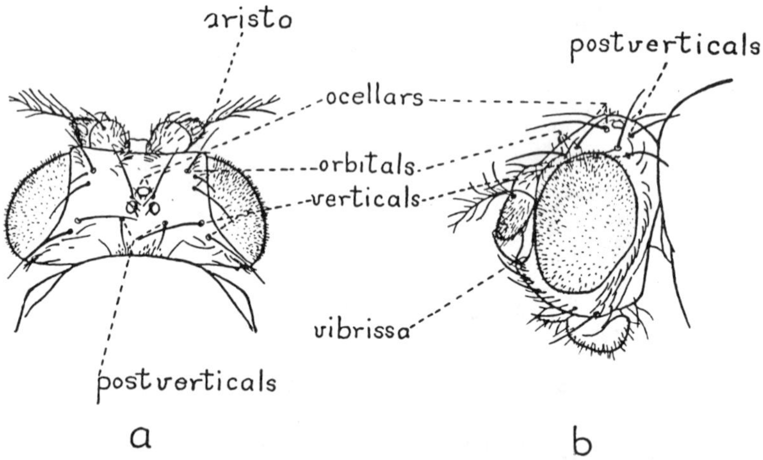

Fig. 2. Top and side views of head of wild-type fly, showing names and positions of main bristles.

When, in 1910, a white-eyed mutant of Drosophila ampelophila (as the species was then called) appeared, there was nothing known at the time to indicate that this species gives rise frequently to mutant types,

[1]) BOGODON was the first, apparently, to use this aseptic method in studying certain physiological problems in the meat fly.

since these had not been seen in the numerous experiments previously made. There were no indications that the white-eyed mutant owed its origin to any peculiarity of the environment, nor anything to show that it had not so arisen. The same statement holds today, for there has been found no evidence of any relation between a specific external influence and a particular mutation. Such a statement is not intended to imply that the change in the germ material that gives rise to a mutant type may not trace back ultimately to an external influence;

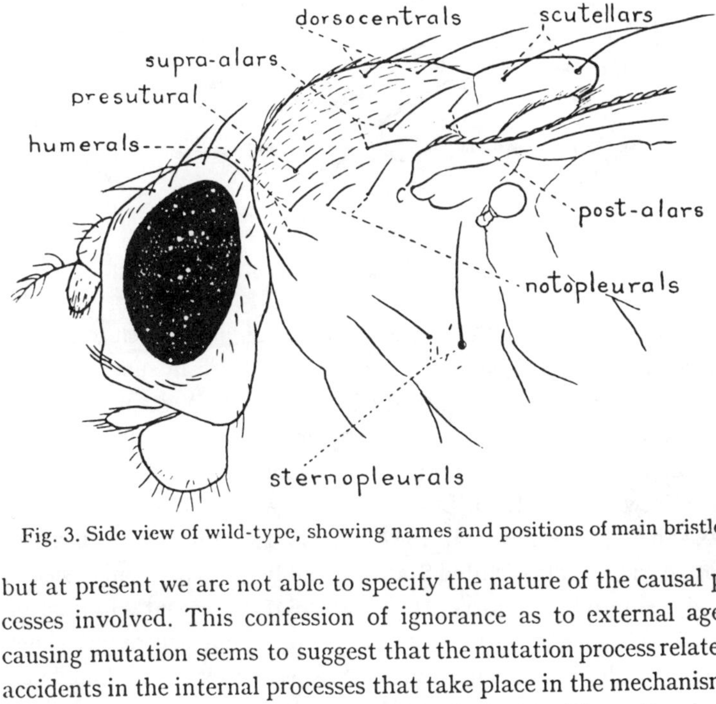

Fig. 3. Side view of wild-type, showing names and positions of main bristles.

but at present we are not able to specify the nature of the causal processes involved. This confession of ignorance as to external agents causing mutation seems to suggest that the mutation process relates to accidents in the internal processes that take place in the mechanism of cell-division or of maturation or of crossing-over. If so, slips in the machine might be so far removed from the sequence of external events from which they ultimately arise, as to escape our power of detection. This possibility can be better discussed in connection with the evidence as to the origin of the mutations of Drosophila and as to the time at which such changes take place in the cells of the germ-track.

The distribution of Drosophila melanogaster is nearly world wide, although it is not found in far northern regions. In the laboratory it is sterile and soon dies if the temperature ranges only a little above 31° C, but in nature it must survive when the temperature of the air goes higher than this, possibly by remaining in moist places near the ground, where, due to evaporation, the temperature may be somewhat lower. The species is common in tropical parts of the world.

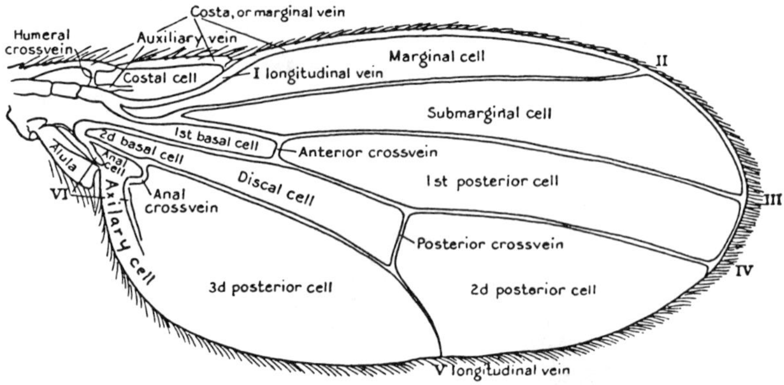

Fig. 4. Wing of wild-type, showing names of veins and cells.

No doubt modern methods of transportation have much to do with local distribution. Fruit stores in the cities furnish in summer one of the best localities for finding it. Curiously enough it was not recorded from North America (Cuba excepted) until 1882, although BOWLES had seen it in 1879 in Montreal. It was recorded later as having been seen in New York in 1875 (LINTNER), in New Haven in 1879 (WILLISTON), and in Massachusetts in 1882 (WILLISTON). Since then it has been reported widely in the United States. Possibly it has been only recently introduced into this country. It is doubtful if it hibernates in the Northern States (outside of fruit and grocery stores) but is introduced anew each year with fruit from the South. The species is primarily found near ripe fruit and in summer is easily obtained by exposing almost any kind of fermenting material. For laboratory purposes, bananas furnish by far the most desirable culture medium, but the fly can also be reared on other fermenting fruits. In our cultures bananas are used, but whether there is something supplied by the banana to the flies that is advantageous, or whether the banana is the best medium for the growth of the yeast, that appears to be the chief element of the food for both the flies and larvae, is not known. It is not improbable that the relative absence of moulds such as mucor, etc. in the acid banana cultures also plays a rôle in the results. How far yeast is a sufficient food in itself will be discussed later.

STURTEVANT ('21) has given a detailed description of Drosophila melanogaster (formerly called ampelophila). The female and male are drawn in Fig. 1 *A* and *B*. The female is ordinarily somewhat larger than the male, the size of both depending on the food conditions of the larvae. The distribution of the bristles on the head is shown in Fig. 2 *a* and *b*. The names and distribution of the bristles on the thorax are shown in Fig. 3. The normal venation is indicated in Fig. 4. The external genitalia of the female and male (ventral view) are shown in Fig. 5.

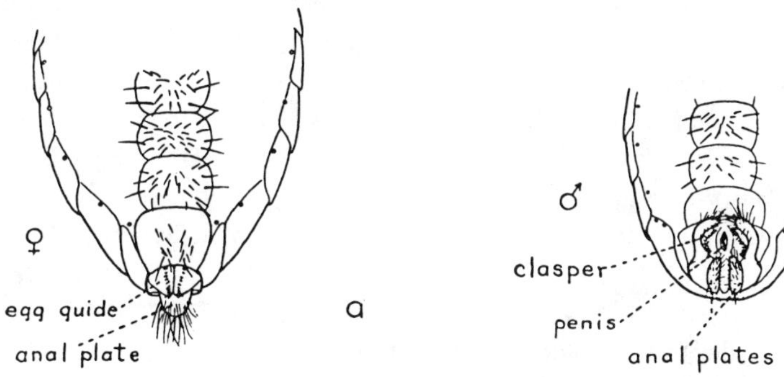

Fig. 5. Ventral views of external genitalia of female and male.

CHAPTER II

BIOLOGY OF DROSOPHILA

The rapid rate of multiplication of the vinegar fly, owing to the large number of offspring produced, and to the brevity of the life cycle, makes it excellent material for genetic purposes. At room temperature or a little above (24 to 25° C), it takes only about 10 days to pass through a complete generation from imago to imago. Under favorable conditions twenty-five generations may be bred in a year.

The length of life of the adult fly has been studied by LUTZ ('11), MOENKHAUS ('11), LIFF ('15), HYDE ('14), PEARL and PARKER ('22, '23, '24), LOEB and NORTHROP ('16, '17). The length of its life has been shown to depend to some extent on the treatment it receives. LUTZ kept a female 80 days and a male 90 days; MOENKHAUS kept a female 153 days; PEARL and PARKER 97 days.

Whether the female continues to produce eggs to the very end of her normal life, and whether the male remains fertile until he dies is not definitely known, but it is probable that both sexes continue to reproduce, almost to the end but, in the case of the female, with greatly diminished productivity.

There have been only a few attempts to find out how many eggs a single female may produce. This is difficult because many of the eggs may fail to develop; in fact HYDE has shown for certain mutant stocks that not more than 50% of the eggs produce flies. It is also known that females of certain of the weaker stocks lay only a few eggs before they die. The females of some mutant stocks are completely sterile; in other races the males are sterile.

The females do not as a rule mate during the first 24 hours. If old males are present, however, early mating is more likely to occur. In fact females with an elongated body (the condition on emergence from the puparium), have been found copulating with old males, but whether the mating was successful is not known. The first eggs are laid usually at the end of the second or during the third day of adult life. The number of eggs laid per day varies greatly with the condition of the food. If the food is unsuited to call forth the egg-laying reaction, the eggs may be retained for several days. Virgin females that have not mated appear to withhold their eggs for a longer period than do females that have mated. If an egg that has passed into the uterus (where the eggs are normally fertilized) is retained there after fertilization it will continue to develop and may even be ready to hatch when laid, or when squeezed out of the female. The other eggs, remaining in the terminal portion of the egg-strings, are not fertilized during the delay in deposition of the uterine egg. In order to obtain eggs for egg-counts or for experimental purposes at regular intervals after fertilization, it is necessary to supply females that have been well fed, with over-ripe bananas or bananas that have been slightly fermented. This method has been described by HUETTNER ('23).

There is a preliminary courtship by the male. If he approaches the female from in front, he circles in half-turns about her with his head kept towards her. One wing may be thrust out for a moment at right angles to his body. Copulation lasts about 20 minutes — the time being a specific characteristic. STURTEVANT ('15*b*) has given an account of the courtship and mating, and has shown that smell rather than vision plays the more important rôle, although vision may take a secondary part in the process. He has also studied the results when two males, belonging to different stocks — one of them usually a mutant stock, — are placed in competition. The results indicate that the more vigorous male is the one more often successful. Special characters present in these males such as eye colors or wing modifications or body colors do not appear to be significant factors in the outcome. On the other hand, when a single male is given two females he more often mates first with the weaker female regardless of her special characters. This is due to the less resistance she offers to his attentions.

The egg of Drosophila melanogaster measures 0.5 mm. long. It is covered with a single layer of six sided cells that can be stripped off — the chorion so-called. Beneath this is the tough chitin coat called the vitelline membrane. Near the anterior end of the egg, there is a small nipple-like protrusion through which a small tube, the micropyle, leads to the surface of the egg. The spermatozoa pass through this tube to penetrate to the protoplasm. More than one sperm may enter each egg. Extending outward from the antero-dorsal surface of the egg are two long spongy protrusions that serve to anchor the egg at the surface of the food. These supporters are formed while the egg is in the ovary out of follicle cells — the same cells that constitute the chorion.

The egg has a distinct bilateral shape as seen in dorsal view (Fig. 6). It has also a dorsal and a ventral side, the difference between them being best seen in side view (Fig. 8, b). The dorsal side is flatter, the ventral is a little more rounded. The micropylar end of the egg corresponds to the anterior end of the embryo and in the oviduct also it lies anteriorly. The posterior end is the first to emerge from the genital opening. It is sometimes said that the dorsal and ventral aspects of the eggs of insects correspond with the dorsal and ventral sides of the mother. However this may be for other insects, the identification of such a relation is not obvious in Drosophila, for, when the eggs have reached the lower part of the egg-strings of an ovary they form a rather compact cluster (Fig. 7a) in which the axes of the egg do not correspond with those of the mother, but the dorsal aspects of the outer eggs lie toward the inside of the bunch, which is nearly spherical in shape. In earlier stages, however, there may be a correspondence, such as that mentioned above, but here again the terminal sections of the egg-strings lie in compact cylindrical form. Nevertheless it is not improbable that each egg-string may have a bilateral structure in common with other organs of the mother, and the formation of the bilateral coat of the egg may be traceable to such a source. Since the bilateral structure of the embryo corresponds with that of the egg, then, if the bilaterality of the envelope of the egg is due to the structrue of the maternal tube, as suggested above, rather than to an inherent structure in the egg itself, it follows that the bilaterality of the embryo is determined by that of the mother in each generation. When the egg reaches the uterus it lies with its axes corresponding to those of the mother.

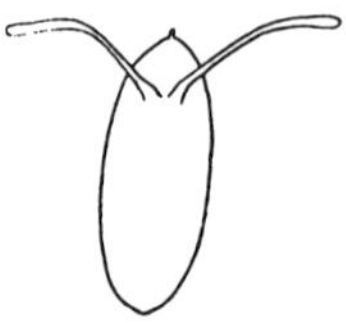

Fig. 6. Dorsal view of egg.

The internal phenomena of reproduction have been studied by NONIDEZ ('20). The external genitalia of the female and male are shown in Fig. 5*a*, *b*. The internal reproductive systems of the female and male are shown in Fig. 7*a*, *b*. The testes are two yellowish tubes; two glandular organs also open into the vas deferens. An ejaculatory sac is present near the lower end of this duct. In the female there are two clusters of egg-strings — the ovaries. The outlet of each ovary is a short oviduct. The two oviducts unite into a single tube which enlarges posteriorly to become the uterus. There are present two mushroom-shaped spermathecae that contain sperm after copulation. Their ducts open into the dorsal wall of the anterior end of the uterus. There is another receptacle for sperm, lying ventrally, with an independent opening into the uterus. This receptacle is a simple coiled tube lying against the uterus (extended in the figure). The sperm from the male are deposited at first in the anterior end of the uterus. They pass into the three receptacles, probably by their own activity, according to NONIDEZ. The ventral receptacle fills up before the spermathecae.

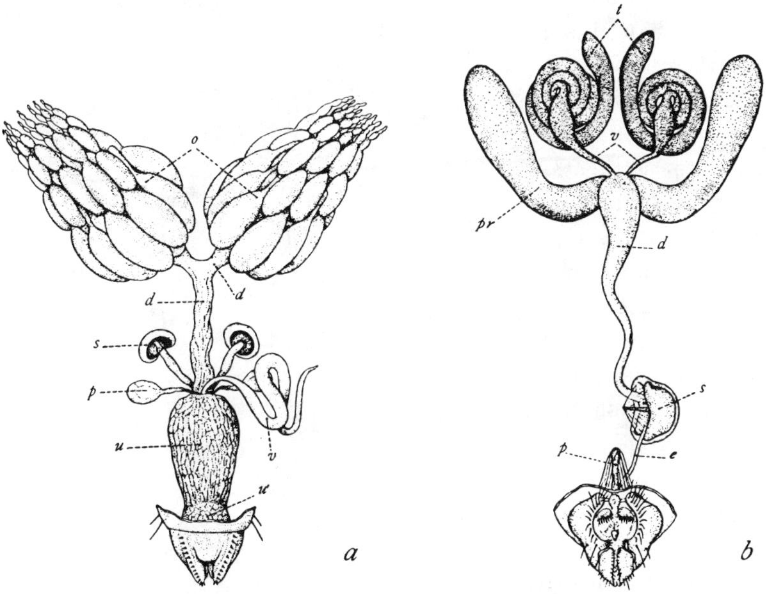

Fig. 7. a, female generative organs, ventral view. d, oviducts; d , azygos oviduct; o, ovaries; p, parovaria; s, spermathecae; u, uterus; u', vaginal portion of the uterus; v, ventral receptacle, represented as uncoiled in this figure. b, male generative organs of Drosophila melanogaster, ventral view, d, vas deferens; e, ejaculatory duct; p, penis; pr. paragonia; s, ejaculatory sac or sperm pump; t, testes; v, vasa efferentia. (After NONIDEZ).

The position of the egg in the uterus is shown in Fig. 8, b. The micropyle (Fig. 8a and b) lies opposite the opening of the ventral receptacle in an advantageous position for receiving the sperm from the ventral tube. Experiments show that the spermatozoa of the ventral receptacle are those first used in fertilizing the eggs. When they have become exhausted the eggs may, NONIDEZ thinks, be fertilized by sperm from the spermathecae, but it is difficult to demonstrate this with certainty. NONIDEZ tried to settle this point by double mating, using a male with a dominant character for the second mating. The result shows, at least, that if the spermatozoa present in the ventral receptacle are used up before the second mating, the first eggs that are subsequently fertilized receive sperm from the se-

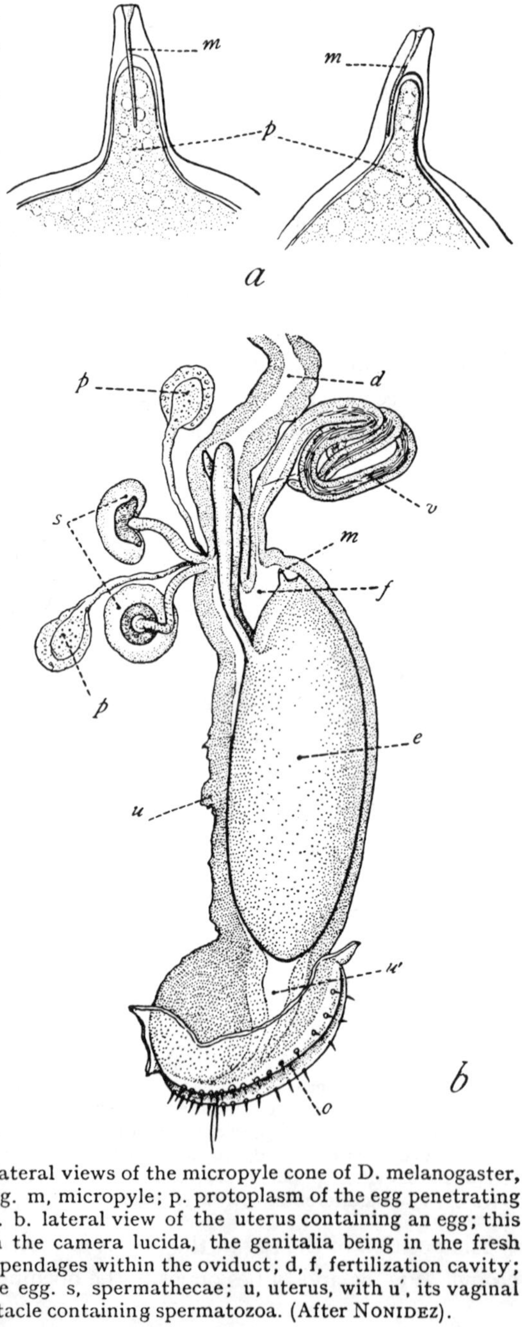

Fig. 8. a, ventral and lateral views of the micropyle cone of D. melanogaster, drawn from the fresh egg. m, micropyle; p. protoplasm of the egg penetrating into the micropyle cone. b. lateral view of the uterus containing an egg; this figure was outlined with the camera lucida, the genitalia being in the fresh condition. e, egg with appendages within the oviduct; d, f, fertilization cavity; m, micropyle cone of the egg. s, spermathecae; u, uterus, with u', its vaginal portion; v, ventral receptacle containing spermatozoa. (After NONIDEZ).

cond male, which are supposedly those in the ventral receptacle. When this is again empty the eggs may then receive sperm of the first male, that may have been kept in reserve in the spermathecae.

The early stages in maturation, fertilization, and cleavage have been examined by HUETTNER ('23). Two polar bodies are formed, or at least two divisons of the egg nucleus give rise to two polar nuclei. The first one divides again so that three polar nuclei and the female pronucleus result. Several sperms enter the egg through the micropyle. Their long tails remain in the protoplasm of the anterior end of the egg. Only one sperm-nucleus unites with the egg nucleus. The other sperm nuclei do not divide, and appear to be absorbed. The segmentation nucleus divides, and the daughter nuclei continue to divide and finally come toward the surface of the egg. When the surface is reached protoplasmic divisions occur, so that each nucleus becomes surrounded by protoplasm which in turn is surrounded by a cell wall. At the posterior end of the egg there is a superficial layer of cytoplasm containing granules that are different from those present elsewhere. They become enclosed in the cells that are formed in this region. Some of these cells later migrate into the interior of the embryo to produce the germ-cells.

In connection with the interpretation of the gonads of gynandromorphs (whose gonads are as a rule either testes or ovaries, even when the right side is male and the left female, and vice versa) it is a matter of interest to discover whether the germ-track is derived from a single cell or from two or more cells that have been formed in the posterior germinal cytoplasm. HUETTNER's evidence has led him to conclude that more than a single cell goes into the development of the germ-track. If the evidence had shown that this track could be traced back to one cell only, a simple explanation could be given of the common result of both gonads being alike in gynandromorphs. This interpretation has, in fact, been suggested by MORGAN and BRIDGES. But since the evidence is otherwise the result must now be interpreted to mean that the chance that the nuclei reaching the germinal region have had a common origin, (i. e. from a particular one of the earlier nuclei) is much greater than that they have arisen from two nuclei of earlier stages.

An account of the characters of the larvae and pupae of Drosophila melanogaster has been given by STURTEVANT ('21b). Except for the tumor-bearing larvae, no mutant types have been found, but this may mean only that the larvae are seldom studied. Likewise, no mutant pupal stages have been observed, although certain mutant stocks have characteristic pupae.

LUTZ ('11) examined the length of life of the larva, pupa and adult fed on ripe bananas and kept at room temperature. One adult female lived 80 days and a male 90 days, but the average length of life of females is given as 26.4 days and of males 32.8 days. LUTZ found that female flies deprived of food, but furnished with water, lived at 20° for 68.6 hours, and males 66.4 hours. At 25° the female averaged 60.7 hours and the male 55.6 hours.

LOEB and NORTHROP ('16, '17) have carried out various experiments with Drosophila melanogaster which bear on the problem of the length of life, not only of the adults but of the larval and pupal stages also. In the earlier experiments ('16) the need for certain food-stuffs was examined. The eggs from bacteria-free flies were placed on dead moist yeast (plus citric acid); the eggs hatched and produced adults whose eggs were treated in the same way. Twelve generations of flies were thus reared aseptically, and yeast was found to supply all that is essential for the life of the fly.

On the other hand when the eggs of bacteria-free flies were put on sterile bananas (or potatoes) the larvae that came from the eggs grew slowly and only a few pupated. Some larvae were, however, still alive after 20 to 30 days. The few flies that went through to adults were small and produced no offspring when mated with each other, or with normally reared flies. NORTHROP ('17) also reported an increased length of larval life due to starvation. LOEB and NORTHROP state that the duration of each of the three stages is independent of that of the other two stages. The conclusion is in contradiction to some observations of PICTET on the caterpillars and pupae of moths in which a definite relation is established between length of larval and pupal life. In the case of Drosophila the length of life of the pupa is so short that unless close readings were made the possible influence of the length of larval life on that of the pupa might be overlooked. In the moth, on the other hand, the larval and pupal life are relatively long and better suited therefore to study this relation.

LOEB and NORTHROP ('17) also determined the length of life of the imago at 25° on agar alone (2 days); on agar plus dextrose and salts (8 days); on glucose-agar (28.5 days). The length of life of aseptic larvae fed on dead yeast and kept at different temperatures (10° to 31.5° C) was studied. The temperature coefficients (Q_{10}) for the larval period show that they are of the order of magnitude of a chemical reaction — namely 2 (or more) for a difference of 10° C. These temperature coeffecients show irregularities, increasing at the lower limit and diminishing at the higher limit. At 10° C, the average duration of the larval period was about 57 days; at 15° C., 18 days; at 20° C., 8 days; at 25° C., 5 days; at 27.5° C., 4 days; at 30° C., 4 days; at 31.5° C., 5 days.

The duration of pupal life was determined by subtracting, from the duration of the time to emergence, the length of larval life. The foregoing data for each temperature were used for the latter period. At 10° C. the pupal life is 13.7 days; at 20° C., 6.3 days; at 25° C., 4.2 days; at 27.5° C., 3.21 days; at 30° C., 3.41 days. When the time of pupal life is plotted on these temperatures it gives a curve that is approximately a straight line.

Adult (aseptic) flies reared on yeast were placed in 500 cc. flasks containing 25 cc. glucose-agar. The time of death of the flies at known temperatures was recorded. Cultures containing 30 to 40 flies had the same average duration of life as those containing 10 to 15 flies (see PEARL). The temperatures used were between 27.5 and 37.4° C. All but the first of these temperatures, it should be noted, are too high for the normal life of this fly. If the temperature coefficients for the duration of life are computed and their reciprocal values are calculated from them and plotted, a slightly curved line is obtained that is more nearly that of a chemical reaction curve than the same curve for larval and pupal life. This is explained as due to the rate not decreasing at the high temperature conditions. It may be questioned, perhaps, whether the rates for the highest temperature do more than show the injurious effect of the high temperatures rather than the natural shortening of the normal life. The authors themselves state that since the adult fly lives less than a week at 5° C, while at 15° it lives 92.14 days on an average, the result "shows that the former temperature is incompatible with the life of the organism." They also state that between 27.5 and 31.5° C. "life is also no longer normal."

This excludes the upper third of the curve. In another paper ('17) the authors sum up their results as follows: "The experiments show that Drosophila has a temperature coefficient for the duration of life of the order of magnitude of that of a chemical reaction. Since the animals used in our experiments were absolutely free from microorganisms, this influence of temperature on the duration of life cannot be attributed to bacterial poisons."

"It was found, moreover, that the duration of the pupal stage is at each temperature proportional to the duration of the larval stage and that the same proportionality exists between the duration of the life of the insect to the larval stage, as far as our present experiments go. Since we know that the duration of the larval stage is determined by a specific hormone we are compelled to consider the possibility that the duration of life is also primarily determined by the formation of poisonous substances or a hormone in the body."

A very complete study of the length of life of Drosophila by PEARL and PARKER ('21, '22) gives quantitative data from which life-curves have been constructed. They made use of wild-type races one of which had been kept under laboratory conditions for several years, and two mutant races also inbred for some time. Pair matings were made and the offspring reared on banana-agar-live-yeast in half-pint milk bottles at a constant temperature (25° C.). The oldest of the wild-type flies lived 97 days, the mutants from 48 to 53 days. The period falls short of that obtained in other cases where by more frequent changes, or by better food conditions, flies have been kept alive for a longer time (MOENKHAUS, LIFF). The mortality curves drawn by PEARL and PARKER show that Drosophila "follows quantitatively the same general law in respect to the distribution of its mortality as does man." This, as they point out, applies only to the imaginal life of the fly. There is no component in their life tables corresponding to the mortality of infancy and childhood in man. It is also clearly shown by their results that the length of life involves hereditary factors, a conclusion that HYDE had reached for one of the mutant races (truncate) and which is also a matter of common observation in our work on mutant races, some of which live for only a few days.

In a second contribution PEARL and PARKER ('22) studied the effect of density of population on the length of life of Drosophila in a confined space. Instead of starting with a pair as above, several or many individuals of the same age and race were put together. In the course of time the flies decreased in number but the number at the beginning is taken as the index of the effect on the length of life. Provisionally, the following conclusions were drawn "(a) the lowest density is not the optimum; (b) the mean duration of life tends to increase with increasing density up to a certain point which is optimum; (c) after the optimum region has been reached increasing density is associated with diminished duration of life which presently falls below the lowest figure found with densities below the optimum."

An even more curious result was obtained by PEARL and PARKER in relation to the number of offspring (adults) obtained when 2 to 50 flies were used as parents in the same bottle for a period of 8 days and again throughout a second period of 8 days in another bottle. "The rate of reproduction per mated female per day declines as density of population increases, at first extremely rapidly and then more and more slowly at higher densities." The authors think that the results can not be due to overcrowding and suggest some other factor is acting, possibly a "psychological" one. It will be necessary to determine the number of fertile eggs laid by the female under these conditions before it will be possible to discover the nature of the factor involved.

CHAPTER III

CHROMOSOMES OF DROSOPHILA

The processes that take place in the germ-track of Drosophila are difficult to study, but by a suitable technique, such as that described by STEVENS ('08), BRIDGES ('16) and METZ ('14) excellent metaphase plates in the oogonial cells and spermatogonial cells have been found. These show that there are four pairs of chromosomes in both sexes. In the female (Fig. 9) there are two rod-shaped or slightly bent X-chromosomes, two large V-shaped autosomes (chromosome III) and two other V-shaped chromosomes (II) nearly as large. In addition there are two very small round chromosomes (IV) that lie as a rule in the middle of the metaphase plate. In the male (Fig. 9) the chromosome group is the same, except that the sex-chromosomes consist of an X, like the X's of the female, and a Y-chromosome that is slightly longer than the X, and is an unequal armed V.

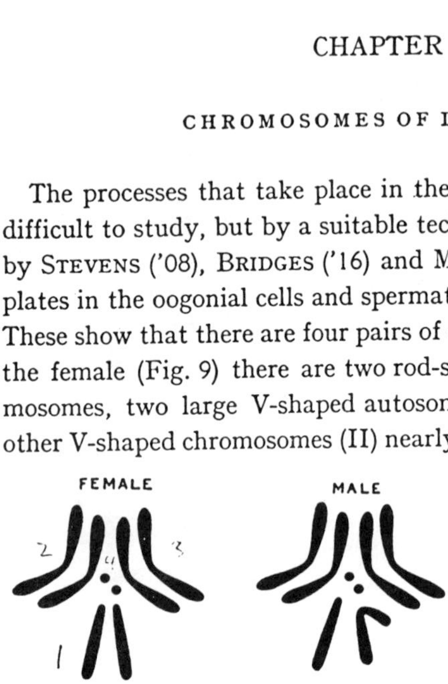

Fig. 9. Diagramatic representations of the chromosomes of the female (left) and of the male (right).

The difference in size between the second and the third pair of chromosomes is not shown in this diagram, but is accurately shown in Fig. 11 a', the pair on the left being the smaller pair, and in Fig. 11b' (the set of three to the left being the smaller). Taking X as 100, the smaller pair of bent chromosomes is 145 long and the larger pair of bent chromosomes is 175. The smallest chromosome (chromosome IV) is 10, the ratio between the arms of the Y-chromosome is 40 to 70.

When the chromosomes, after splitting, move towards the pole, i has been shown that the spindle fibres are attached to the middle of the two large autosomes (where they are bent) and the inner end of the X, and at the bend of the J-shaped Y-chromosome.

The first account of the chromosomes of Drosophila melanogaster was that of Miss STEVENS ('08). This account was essentially correct, but owing to a wrong interpretation of the sex-chromosomes, it gave rise at first to an erroneous conclusion in regard to an important point. Her three figures of oogonial metaphase plates (Figs. 80, 81, 82) were correct; some of the spermatogonial figures were nearly correct (Figs. 58, 59, 60) but the other figures for the male, that represent the chromosomes about ready to divide, were erroneously interpreted, due apparently to a mistake in the interpretation of the X-chromosome. This was represented as Y-shaped. One branch of the Y was supposed to represent the X-chromosome attached to another V-shaped chromosome. By implication there would be two small X-chromosomes in the female attached to two other chromosomes (called XX above). This branched chromosome is probably due to the precocious division of the long arm of the bent Y-chromosome of the male.

Following STEVENS' conclusion the earlier genetic work of Drosophila assumed that the formula for the male was XO instead of XY. As a matter of fact this assumption made no difference so far as the essential relations were concerned, since it is now apparent that the Y-chromosome of Drosophila carries no factors that affect the influence of recessive factors carried by the X-chromosomes. Hence for genetic purposes the Y can be treated as non-existent.

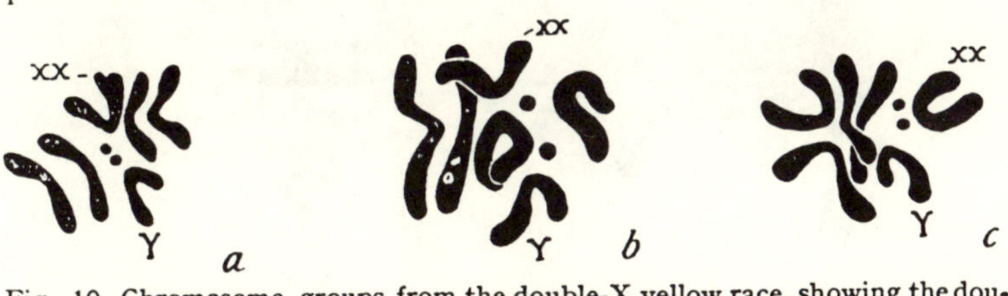

Fig. 10. Chromosome groups from the double-X yellow race, showing the double-X as a large V, and the J-shaped Y-chromosome.

METZ ('14) gave two figures of the female of Drosophila melanogaster that are practically identical with STEVENS' figures for the female. In the same year BRIDGES ('14) found that the female Drosophila melanogaster has two X-chromosomes and the male an unequal XY pair — a statement based on cytological preparations. He also showed that the non-disjunctional females carry two X's and a Y-chromosome. Later

BRIDGES pointed out that the Y-chromosome is not a straight rod, but is a rod that is sharply bent near one end. Later work has confirmed this identification (BRIDGES '16).

The most favorable age for obtaining oogonial divisions is that of flies that have just emerged or are in later pupal stages. Aceto-carmine preparations, which are not permanent, often give fine pictures. Ovaries fixed in strong Flemming solution, that gives excellent results, furnish the standard material for most studies, and give verification of the results obtained by the former method. The nuclei during the stages at which the conjugation of the chromosomes takes place are large, clear, and lightly staining, and very difficult to interpret. As yet no information relating to crossing-over has been found.

The spermatogonial divisions are shown (by either method) in the cells near the anterior end of the testes, but they are less often observed than the corresponding stages in the female. Little or nothing is known of the synapsis stages.

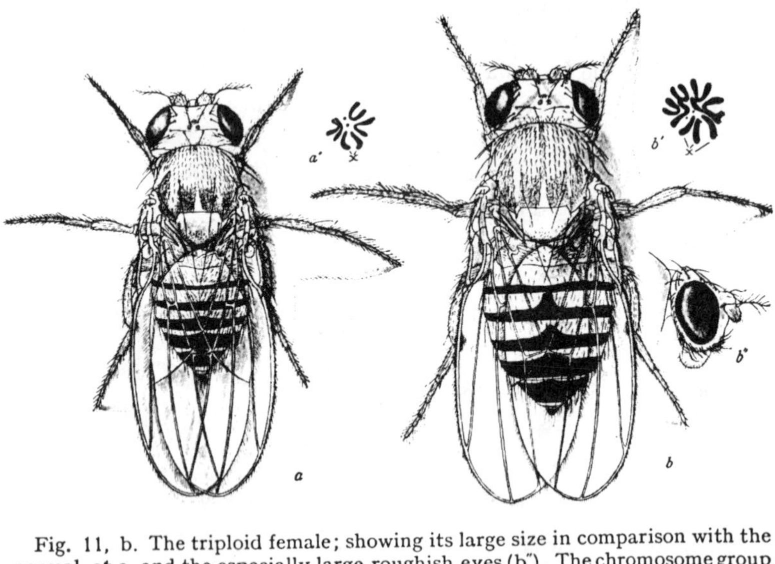

Fig. 11, b. The triploid female; showing its large size in comparison with the normal, at a, and the especially large roughish eyes (b''). The chromosome group (b') consists of three of every chromosome.

Several of the aberrant genetic types have been shown to be connected with changes in the number of the chromosomes. Females that show secondary non-disjunction have been shown by cytological preparations to contain a Y-chromosome (BRIDGES '16).

A peculiar type of female that reverses the ordinary type of criss-cross inheritance has been shown (L. V. MORGAN '23) to have its two X-chromosomes united at one end (Fig. 10), so that they behave as a unit in the reduction division. Triploid females have been discovered (BRIDGES '21) and their chromosome group shown to consist of three members of each kind of chromosome (Fig. 11). Tetraploids have been also found (BRIDGES and L. V. MORGAN) but as yet their chromosomes have not been studied. Haplo-IV and Triplo-IV flies have been shown (BRIDGES '22) to have one and three IV-chromosomes respectively (Fig. 12a, b).

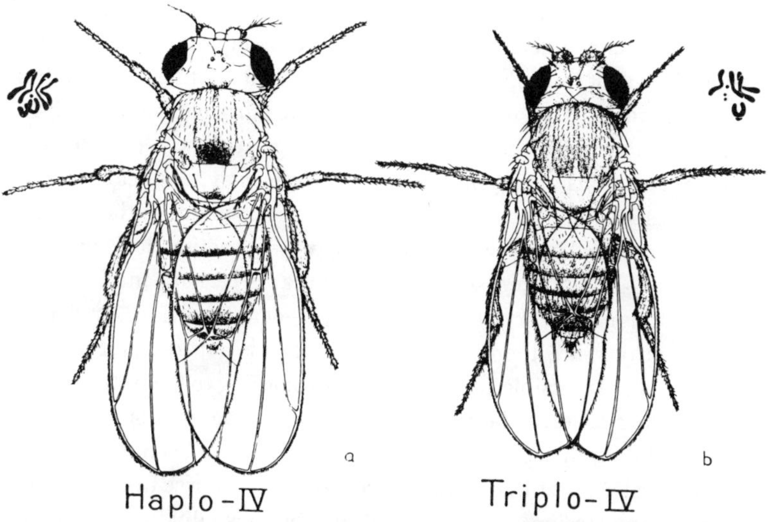

Fig. 12. The haplo-IV female (a) contrasted with the triplo-IV female (b) with their respective chromosome groups.

Measurements of the chromosomes and of the sperm head have been made in order to obtain data for calculations relating to the upper and lower limits of the possible size of the genes (MORGAN '22). These speculations are dealt with on pages 85—86.

The chromosome groups of other species are described in Chapter XVIII.

CHAPTER IV

MUTATION

It is usual to define mutation as a transmissible change in the germ-plasm, producing a detectable change in the characteristics of the individual. If this definition is accepted it is evident that further subdivision is desirable. Two main types are at once distinguishable: (a) increase or decrease of whole chromosomes (e. g., triploidy, non-disjunction) or of sections of chrcmosomes (e. g., deficiency, duplication), and (b) changes at single loci. The first type is discussed under the heading of "Chromosome Aberrations", while the present section deals with the second type and with a case (that of bar) that is perhaps intermediate in nature. It is not improbable that the single-locus mutations, or, better, gene-mutations, are themselves of quite diverse natures. But since at present it is not possible to classify them according to the nature of the process occurring, it will be necessary to treat them here as a single group.

Several hundred separate cases of mutation have been detected in the various species of Drosophila, and this has given rise in some quarters to the impression that mutations must be unusually frequent in the group. The facts do not seem to warrant this conclusion. It may be estimated that at least 13,000,000 (probably nearer 20,000,000) flies have been etherized and examined by competent observers. And practically all these have been from thoroughly known sources, so that any new types would be recognizable as unexpected and would have an excellent chance of being detected and tested. The actual frequency of mutation can scarcely be determined, even roughly, partly because it is so low as to present serious statistical difficulties, and even more because detection of mutations is so largely dependent on the personal equation of the observer.

Muller and Altenburg ('19) have examined the frequency of occurrence of new lethal mutations in the X-chromosome.

They tested 1062 females, and found 20 new lethals among them, or a ratio of 1 in 53. Since each female carries two X-chromosomes, this means, if the numbers are significant, that about 1 per cent of the X-chromosomes in a population carry newly arisen lethals. A similar experiment with the second-chromosome gave about 30 mutations in 5000 chances (MULLER '23).

The specific mutation frequency for any given gene is so low that it has not been found possible to get even an approximate measure of it except in the case of the bar allelomorphs (to be discussed below), though MULLER ('23) has reported two genes that gave, respectively, 4 lethal mutations in 5,000 gametes and 2 in 1800 gametes. Further study may show these two loci to be favorable for quantitative determinations of the frequency of mutation.

It is clear that mutation frequency is not identical for all genes. Certain loci have been observed to mutate repeatedly, while others that give equally conspicuous mutant characters have not been found to mutate more than once. Table 1, showing the loci that have mutated more than once, will illustrate this point.

TABLE 1. — RECURRENT MUTATIONS AND ALLELOMORPHIC SERIES

Locus	Total Occurrences	Distinct mutant Types	Locus	Total Occurrences	Distinct mutant Types
apterous. . . .	3	1	cut	16 +	5 +
ascute.	4 ±	1	dachs	2	2
Bar.	2 +	2 +	dachsoid . . .	2	1
bent	2 +	2	Delta	2	2
bifid	3	1	deltex.	2	1
bithorax. . . .	3	2	Dichaete . . .	3	3
black	3 +	1	dusky.	6 +	3
bobbed	6 +	1	ebony.	10	5
brown.	2	2	eyeless	3	2
broad	6	4	fat	2	2
cinnabar . . .	4	3	forked	12	4
club	2	2	fringed	2	1
crossveinless . .	2	1	furrowed . . .	2	2
curved	2	2	fused	2	2

Locus	Total Occurrences	Distinct mutant Types	Locus	Total Occurrences	Distinct mutant Types
garnet	5	3	ruby	6	2
Haplo-IV . . .	35 ±	1	rudimentary . .	15 +	5 +
inflated	2	1	sable	3	2
lethal-a	2	1	scarlet	2	1
lethal-b	2	1	scute	4	1
lethal-c.	2	1	sepia	5	1
lethal-e	4	1	singed.	5	3
Lobe	6	3	Star	2	1
lozenge	10	5	tan	3	2
maroon	4	1	tetraploidy . .	3	1
miniature . . .	7	1	triploidy . . .	15 ±	1
Notch	25 ±	3	Truncate . . .	8 ±	5
pink	11 +	5	vermilion . . .	15 ±	2
purple	6	2	vestigial . . .	7	5
reduced	2	2	white	25 ±	11
rough	2	2	yellow	15 ±	2
roughoid . . .	2	2			

The numerous mutations found in Drosophila furnish material for an investigation of the stage at which mutations occur. Since dominants and sex-linked recessives in males, produce visible effects at once, whereas autosomal recessives do not appear until after the mating of two heterozygous individuals, the two former classes of mutations furnish the clearest evidence on the question. Bridges ('19*e*) estimated that approximately half of the sex-linked recessives of melanogaster for which critical data are available, were first observed each in a single male, and that the majority of the remaining ones first appeared as half the sons from single mothers, many of these mothers evidently having received the mutant gene from a parent. This evidence indicates that at least half of the mutations occurred at, or not many cell-generations before, the maturation stages. This conclusion is also in agreement with the fact, reported by Bridges, that fully two-thirds of the then known dominants were first found as single heterozygous individuals. Bridges also reports several cases in which mutation must have occurred at some point in the history of the germ-cells after fertilization and several cell-generations before maturation. Numerous other such cases are on record (e. g., MULLER '20*b*, LANCEFIELD '22, MOHR '23*a*, etc.). MULLER, in the paper cited, has analyzed the process of deducing the stage at which mutation occurs from the numbers of individuals first appearing with the new character. He concludes that, if the mutation process is as likely to occur at any one stage in the germ-track as at any other stage, then by far the most frequent result will be that actually observed — i. e., that the new form will appear in a single individual. Mutation is not limited to any one stage in the life-cycle, but whether or not it is more frequent at some stages than at others is not demonstrated.

In several cases (MORGAN and BRIDGES '19, MULLER '20*b*, STURTEVANT '21*a*, MOHR '23*a*, and unpublished data) somatic mosaics have been found that were evidently due to mutations occurring after the first cleavage division. In consequence part but not all of the body carried the new gene and showed the character determined by it. In MOHR'S case and in an unpublished case in Drosophila simulans, both the soma and the germ-cells received both mutated and unmutated genes; and in MULLER'S two cases some, at least, of the germ-cells received the mutated gene. In these last four cases it was thus possible to demonstrate that a new sex-linked recessive had been produced by a mutation occurring early in the cleavage stages of the developing individual.

The great majority of the mosaics that are to be interpreted as due to mutation rather than to elimination of a chromosome or to double nucleus eggs (see chapter XIV) have apparently been due to the occurrence of new sex-linked recessives, or, more rarely, of new dominants. The four discussed in the last paragraph, that were proven to be of the former type (sex-linked recessives), were all males, — and mutational mosaics that have not been analyzable genetically (because of failure of the mutation to affect the germ-cells) have been males far more often than females. Evidently mutation commonly affects only one of the two homologous chromosomes in an individual; the new gene can accordingly produce no visible effect on the soma unless it is dominant, or is sex-linked and is in a male, where there is no homologous X to carry a dominant wild-type allelomorph.

The fact that mutations commonly occur in only one of the two homologous chromosomes in a given cell makes the possibility of inducing them by altering the environment seem remote; for it would be difficult to imagine conditions more nearly alike than those surrounding corresponding loci in like chromosomes in a single cell. As a matter of fact, none of the numerous attempts to induce mutations in Drosophila have been successful. It may be useful to summarize here the literature dealing with these negative results.

MORGAN ('15c) showed that the dominant character abnormal abdomen is strongly influenced by the environment. It is most marked in flies reared under moist conditions, and may be suppressed entirely under dry conditoins. But rearing the stock for several generations under dry conditions did not at all decrease the intensity of the character in the offspring when reared under moist conditions. Similar failure to transmit characters induced in the parents by environmental agencies has been shown by HOGE ('15a) for reduplicated legs, by PLOUGH ('17b) for modified crossover values and by KRAFKA ('20a) for ommatidia number in Bar eyes — all induced by temperature changes.

PAYNE ('11) bred sixty-nine generations of flies in the dark. At the end of the experiment the flies of the 69th generation were tested for their reaction to light. They showed no significant differences from the controls.

DUNCAN ('15b) conducted an extensive experiment in an attempt to induce mutations by crossing races from widely separated localities. No increase in mutation frequency was obtained — a result that agrees with the general experience of students of Drosophila (see STURTEVANT '21b).

It has often been supposed that allelomorphic genes influence, or "contaminate" each other in heterozygotes. This possibility has been excluded in those cases where it has been looked for in Drosophila — e. g. in dachs (MULLER '16), in curved and balloon (MARSHALL and MULLER '17), in Beaded (MULLER '18), in Trunctae (ALTENBURG and MULLER '20), in Dichaete (STURTEVANT '18a), and in Notch (MORGAN '19a).

MORGAN ('10a, '14a) used "wide ranges of temperature, salts, sugars, acids, alkalis, without any resulting mutation".Some mutations appeared after treatment with radium (MORGAN '11c), but repetition of the treatment on a larger scale, and use of X-rays, both failed to give mutations (MORGAN '14d, GUYÉNOT '14, and more recent unpublished data). LOEB and BANCROFT ('11) also observed a few mutant types after treatment with radium, but MORGAN ('14d, p. 709) has shown that these mutant types were probably present in their stocks before the treatment was applied.

GUYÉNOT ('14) used high temperature, radium, X-rays, and ultra-violet in attempts to produce mutations. The results were negative, except that from the ultra-violet series he obtained in one lot a dark body-color, apparently a recessive mutant type. Repetition of the ultra-violet experiment has failed to produce any mutations (unpublished data).

All the experimental cultures are normally from parents that have been etherized at least once. MORGAN ('14d) tested the hypothesis that this may be the cause for the mutations that occur, by increasing the ether treatment to a dose that was almost lethal. No mutations were found among the 31,168 F_1 and F_2 flies from treated individuals.

MAVOR ('21a, '21b, '22, '23a) has shown that treatment with X-rays causes non-disjunction of the X-chromosomes. In this way XXY females are produced, and these in turn produce further non-disjunction. MAVOR remarks that "there is therefore in this effect of X-rays on the germ-cells a very clear case of an external agent which modifies the mechanism of inheritance in such a way that a permanent effect is produced on the germ-cells." The case is discussed in the section on "Chromosome Aberrations", — we need point out here only that MAVOR has reported no gene mutations whatever following treatment with X-rays.

MANN ('23b) has treated D. melanogaster with arsenic, methyl alcohol, ethyl alcohol, quinine, morphine, strychnine, copper sulphate, lithium carbonate, lead acetate, and methylene blue. She also used high and low temperature (31.5° C. and 2.5° C.), and accumulation of the gases resulting from fermentation of the culture medium. In extensive experiments she obtained actually a smaller proportion of mutants (1 in 23,333) than in the controls (1 in 15,165). High temperature yielded one clear mutation (apricot), but repetition of the experiment with the same stock failed to produce any further mutations, and the extensive experiments of NORTHROP ('20), PLOUGH ('17b), KRAFKA ('20a), PLOUGH and STRAUSS ('23), and others also show that this temperature does not cause any clear increase in mutation frequency.

MULLER and ALTENBURG ('19) have suggested that high temperature does increase the frequency of occurrence of lethal mutations, — but their data are clearly not statistically significant, so that the question of the effect of temperature on mutation frequency must be kept open for the present.

Table 1, page 23, shows that certain loci (e. g., white) have produced several distinct allelomorphs, while others (e. g., miniature) have mutated frequently but appear to have produced always the same new gene. Evidently, then, some genes are unstable in several directions, while other genes are relatively much more unstable in one direction than in others. The familiar "Galton polygon" is at once suggested by these relations, and may probably be taken as a useful analogy.

The series of allelomorphs of white is also interesting in that in several cases mutant allelomorphs have themselves undergone further mutation. Eosin arose from white (MORGAN '12a), and what appears to be the same gene has also come from wild-type. Eosin (derived from one or the other of these sources) has likewise reverted to wild-type and has given rise to white again. This phenomenon of reversion to the wild-type allelomorph has also been observed very rarely in other mutant genes (e. g., scute and black).

It is a striking fact that multiple allelomorphs nearly always affect the same somatic characters. The various allelomorphs of white all affect the color of the eyes and of the envelope of the testes, those of ebony the color of the body, those of forked the form of the bristles and hairs. It seems probable that exceptions to this rule will occur; and cases that are partial exceptions are already on record. Thus, while the three singed allelomorphs all affect the form of the bristles and hairs, in two the females are fertile while in the other the females are sterile; crooked affects the bristles markedly and the eye-surface slightly, while its allelomorph, furrowed, affects the eye-surface markedly and the bristles slightly. It is evident that if this differentiation were carried only a little further we should have cases where allelomorphic genes did not affect any character in common. It is possible that this occurs in the case of some of the genes allelomorphic to Truncate (chromosome II); but this conclusion must be tentative until further data are available concerning the possibility that some of the mutant types here are deficiencies rather than single-gene mutations.

In the preceding discussion of mutation we have avoided considering the bar allelomorphs, reserving them for separate treatment. There are more data at hand for a consideration of this case than for any of the others; but it seems probable that we are here dealing with a special case, that represents a process different in kind from most gene mutations.

TICE ('14) found a single male with narrow eyes (Fig. 13b). The new type proved to represent a new sex-linked dominant character, now

FIG. 13. Bar eye (b) contrasted with the wild-type (a) and ultra-bar or double-bar (c).

known as bar, with a locus near the right end of the X-chromosome map. This is the only case, so far, in which the normal allelomorph of bar ("round") has ever been found to mutate in melanogaster, and there are no mutant characters suggesting bar in any of the other species. It must, therefore, be supposed that this round allelomorph is among the more stable genes of Drosophila. But the mutant allelomorph bar has turned out to be by far the most mutable one yet discovered in that animal. It was first found by MAY ('17) that bar reverts to round; and ZELENY ('21a) has shown that such reversions occur in about 1 in 1600 of the offspring from bar stocks. ZELENY ('19, '21a) discovered another more extreme allelomorph called ultra-bar; and showed that it is produced by bar stocks in about 1 in 28,000 individuals. Further, ZELENY showed that ultra-bar gives round in about 1 in 1700 individuals, and gives bar in about 1 in 2900. ZELENY ('23) further showed that the reverted individuals from a homozygous bar stock are heterozygous females and normal-eyed males in about equal numbers; he concluded from this that bar reversion occurs chiefly or only, in the germ-cells of females. STURTEVANT and MORGAN ('23) verified this result, by showing that reversion does not occur in bar males. They also showed that bar ultra-bar heterozygotes produce reverted ("round") gametes. The most significant contribution of STURTEVANT and MORGAN was to show that reversion of bar is correlated with crossing-over at, or very near, the bar locus. Females were used that were homozygous for bar and were heterozygous for forked ($^1/_5$ unit to the left of bar) and for fused (2 units to the right of bar). They reported six reversions from such females, and all six also carried X-chromosomes that had undergone crossing over between forked and fused. Such crossovers include less than 3% of the eggs from such mothers — yet they included all six of the reversions. Further unpublished data show the same result on a much larger scale. A few reversions have apparently been produced that did not carry forked-fused crossover X-chromosomes; but these are extremely rare, and it is still just possible that they are all due to experimental errors. In any case, it is now clear that the great majority of the rounds and also of the ultra-bars produced by bar stocks carry X-chromosmes that have crossed over between forked and fused. Furthermore bar ultra-bar heterozygotes produce rounds (as shown by STURTEVANT and MORGAN), and ultra-bar round heterozygotes produce bar — these two events again being associated with forked-fused crossing-over.

These results strongly suggest that the mutations in this locus (with the probable exception of the original mutation from round to bar) are due to unequal crossing-over. If we represent a homozygous bar female as $\frac{ABC}{A'B'C'}$, where B represents the bar gene and A and C represent neighboring genes, it seems likely that crossovers ABB'C' and A'C (or A'B'BC and AC') are produced. On this view BB represents ultra-bar, and "reverted" round is to be interpreted as the absence of B.

In one instance bar has given rise to a new allelomorph that is intermediate between bar and round — known as infra-bar. This event occurred in the germ-cells of a male. Because of this fact and of the later behavior of infra-bar, it is probable that this change was not in any way connected with crossing-over, and that infra-bar represents a qualitative, rather than a quantitative, change in the bar gene. Since infra-bar shows the same unequal crossing-over behavior as does bar itself, it has been possible through its use to make a rigorous test of the hypothesis of unequal crossing-over.

Females homozygous for infra-bar behave like bar females in that they give rise to rounds and also to a new type analogous to ultra-bar, again only when forked-fused crossing-over occurs. The new type, called double-infra-bar, has been tested in heterozygotes with round. Such females give infra-bar again, in the same way as ultra-bar over round gives rise to bar.

Females heterozygous both for bar and for infra-bar as well as for forked and for fused $\left(\frac{\mathrm{f\ B^1}}{\mathrm{B\ fu}}\right)$ have produced round and also a new type, BB^1 (not forked or fused). This new type, which is only very slightly less extreme than ultra-bar in appearance, has been tested against round. It gives rise both to bar and to infra-bar, with a frequency (about 1 in 1000) that is apparently the same as that with which ultra-bar over round or double infra-bar over round give rise to the single types. That is to say, the bar and the infra-bar maintain their identity in the double form and can be recovered from it. In addition, by following forked and fused it has been possible to show that the components of bar-infra-bar maintain not only their identity but also their sequence. For example: $\frac{\mathrm{BB^1}}{\mathrm{f\ fu}}$ gives rise to forked infra-bar and to bar fused—never to forked bar or to infra-bar fused. It will be observed that this is simply the usual test for sequence of non-allelomorphic genes. It should therefore be possible to obtain another type B^1B, like the one just discussed except with the sequence reversed. This has in fact been done: $\frac{\mathrm{B}}{\mathrm{f\ B^1\ fu}}$ produced forked infra-bar-bar. And this, when tested against round, gave both bar and infra-bar; furthermore the relation of these to forked and to fused showed that the sequence actually is B^1B.

ZELENY and others have shown that the number of facets in bar eyes is subject to changes due both to modifying genes and to temperature (see section on multiple factors and selection). In order to carry out a series of facet counts on the various combinations of the bar allelomorphs described it was therefore desirable first to obtain a closely inbred stock. Such a stock, apparently uniform for all modifying genes, has now been thoroughly tested. All the counts were made on flies reared at a temperature very close to 25° C. The following table shows the average facet numbers of various types produced in this way.

TABLE 2

Females heterozygous for		Ultra-bar	Bar-infra-bar	Double-infra-bar	Bar	Infra-bar	Round
	Ultra-bar	25.0	24.1	26.7	36.4	41.8	45.4
	Bar-infra-bar	24.1	26.7	27.9	37.0	37.8	50.5
	Double-infra-bar	26.7	27.9	38.2	38.3	138±	200.2
	Bar	36.4	37.0	38.3	68.1	73.5	358.4
	Infra-bar	41.8	37.8	138±	73.5	320.4	716.4
	Round	45.4	50.5	200.2	358.4	716.4	779.4
	Males	29.0	29.7	46.0	84.4	478.1	738.8

Full discussion of these facet counts will be given in future publications; only one relation need be pointed out here. In every case where the point can be tested *two genes in the same chromosome produce more effect than do the same two genes when they lie in homologous chromosomes.* The following comparisons are available:

$$\frac{B}{B} = 68.1 \pm 1.1 \qquad \frac{BB}{+} = 45.4 \pm .2$$

$$\frac{B}{B^i} = 73.5 \pm 1.2 \qquad \frac{BB^i}{+} = 50.5 \pm .4$$

$$\frac{B^i}{B^i} = 320.4 \pm 14.2 \qquad \frac{B^iB^i}{+} = 200.2 \pm 8.6$$

Evidently in all three cases there is an increase in effectiveness when the two genes lie in the same chromosome. The results could be accounted for without assuming any "position effect", if it could be supposed that the round (= " + ") allelomorph has an effect on facet number in the same direction as that of the bar allelomorph. But in this case, reverted bar should lack the round allelomorph, and BB over reverted-bar should resemble homozygous bar. As a matter of fact, a reverted bar and a reverted infra-bar have both been tested, and in both cases BB over reverted was, if anything, slightly *smaller* than $\frac{BB}{+}$. As pointed out in the section on chromosome aberrations, there is a possible analogy to this position effect in the dominance relations of mutant genes in duplications and in triploids.

MULLER ('23) has reported numerous lethal mutations that arose in females in which crossing over was prevented by special genes. That mutation may occur in somatic cells or in males is known (see above), and crossing over does not commonly occur in either of these conditions. It seems almost certain, therefore, that many mutations occur without any relation to crossing over. Unpublished data are also available from an experiment in which numerous double-crossover X-chromosomes were tested to see if they carried new lethals. No mutations at all were found in these chromosomes — a result that again indicates the behavior of the bar series of mutations to be exceptional.

The interpretation of the bar case that has been given is, obviously, a quantitative one. Bar reversion has, moreover, been treated as a "loss-mutation" in the strictest sense of that term. The arguments in the last paragraph, together with the exceptionally high mutation-frequency of the bar series of mutant allelomorphs, lead us to conclude that such losses and quantitative variations in genes are not the basis of most mutations. In fact, it seems probable, even in the bar locus, that the round-to-bar and bar-to-infra-bar mutations were qualitative changes.

CHAPTER V

MULTIPLE ALLELOMORPHS

The occurrence of a pair of genes that disjoin at the maturation stage is the most frequent situation in genetics, but as more mutant types have become known an increasing number of cases have appeared in which there may be more than two genes that act as allelomorphs to each other, when any two are brought together. When three or more allelomorphs are known only two of them can be brought together in any 2N individual. Thus if three genes a, a^1, a^2 are allelomorphic, the possible combinations in diploid individuals are aa, aa^1, aa^2, a^1a^1, a^1a^2 a^2a^2.

In all cases of multiple allelomorphs it has been found that the genes affect the same character or characters in approximately the same way. Thus, in Drosophila the eleven genes at the white locus all affect the eye color, diluting it to various degrees. The numerous genes at the cut locus all affect the outline of the wing (Fig. 14); those at the vestigial locus the size of the wing (Fig. 15). The genes at the forked locus affect the spines, making them twisted or forked.

When a member of a series affects more than one organ (such as the white eye gene which affects also the color of the testes) the other members of the series also affect the same part to some degree.

Another relation is also significant. When two of the genes other than the wild-type gene from which they are derived, are brought together, they do not give the wild-type character but a character that is as a rule more or less intermediate between the two. Thus, when eosin and white are combined (MORGAN and BRIDGES '13) the eye color is somewhat lighter than eosin. Cases may arise, of course, where one member of the series so far dominates the other when in combination that the dominance may appear complete or nearly so. The two allelomorphic recessive genes do not produce a character in excess of that produced by the more effective gene. These relations do not differ in any essential respect from the ordinary MENDELIAN relations of domi-

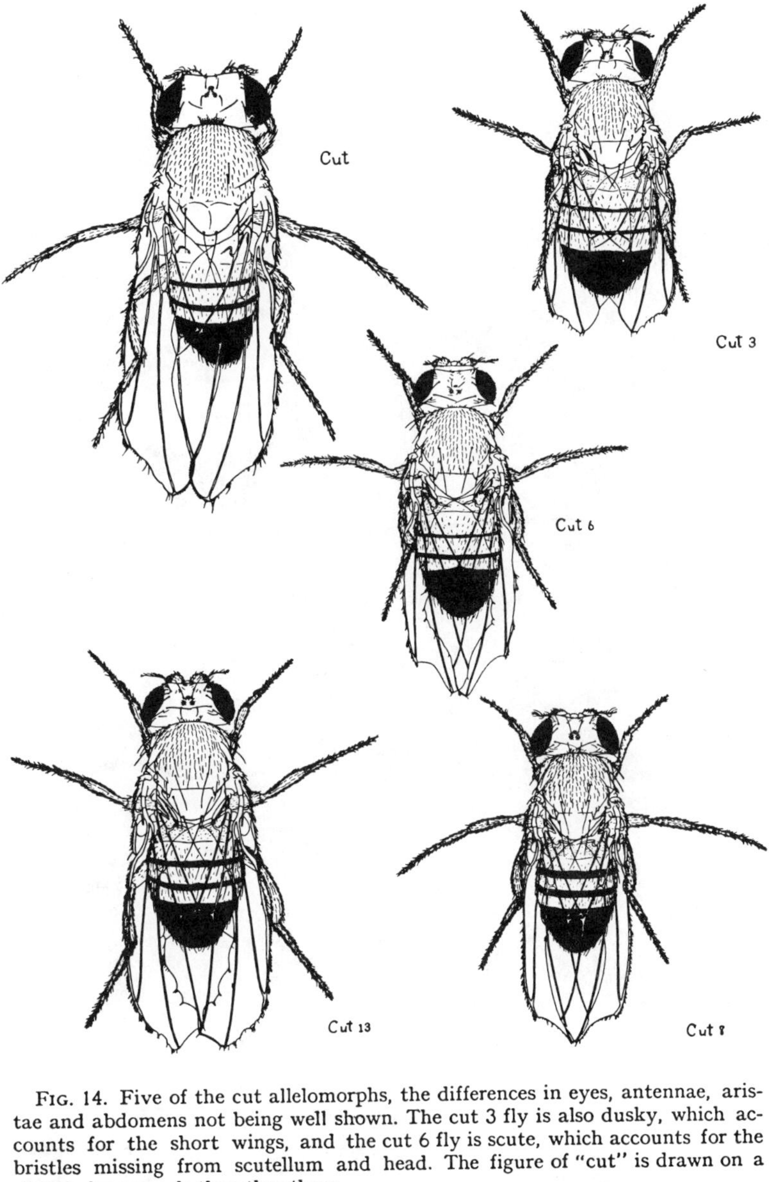

FIG. 14. Five of the cut allelomorphs, the differences in eyes, antennae, aristae and abdomens not being well shown. The cut 3 fly is also dusky, which accounts for the short wings, and the cut 6 fly is scute, which accounts for the bristles missing from scutellum and head. The figure of "cut" is drawn on a slightly larger scale than the others.

nance and recessiveness, although on the whole, the "compound" is more strikingly intermediate. This may be interpreted to mean only that the difference in effect between such recessive genes is as a rule less than that found between the wild-type gene and its derived recessive. In

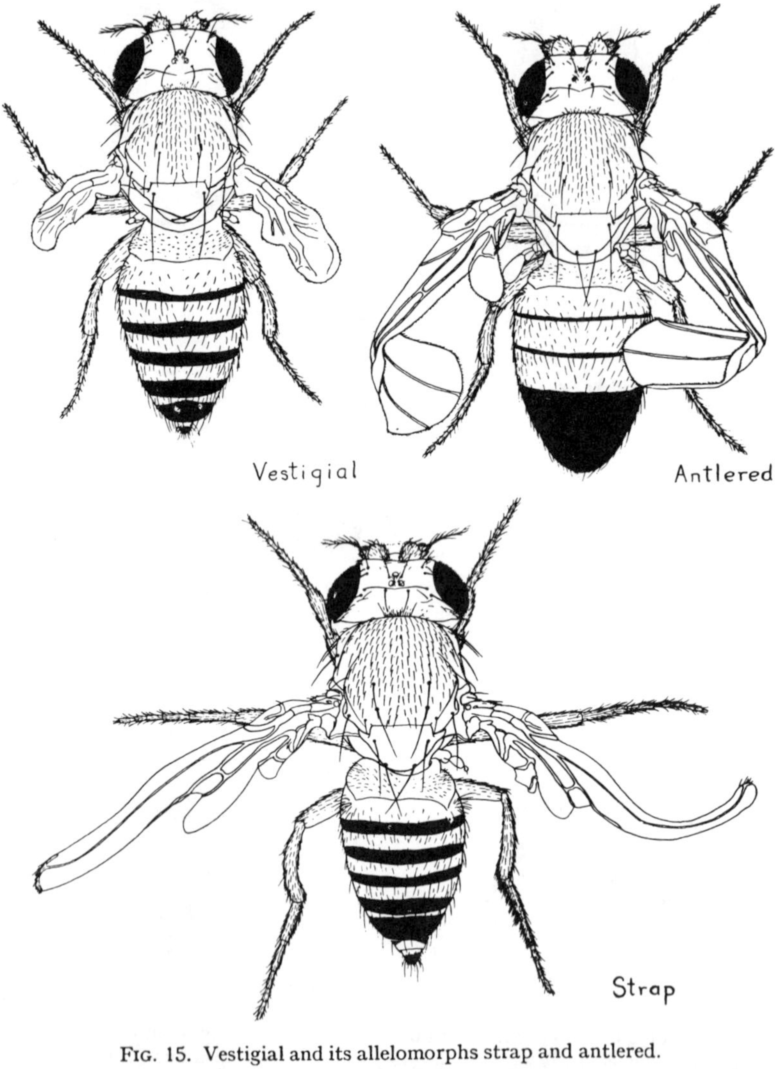

FIG. 15. Vestigial and its allelomorphs strap and antlered.

the latter case the influence of the recessive in the heterozygote is often too slight to be readily detected, hence dominance is said to be complete.

The possibility of the occurrence of more than two kinds of gene at a given locus raises several questions concerning the nature of the gene itself. A popular interpretation of the two members of a pair of allelomorphs is that the dominant represents a presence of something and a recessive its absence. Taken in its original intention, this picture of the relation of the two genes in question is rendered absurd by the discovery that there may be more than one "absence" — a whole series of eleven "absences" at the white locus. But if the definition is not taken literally then the following considerations have a bearing on the question. First. It might be claimed that the phenomenon of multiple allelomorphism is an expression, not of changes in the same locus, but changes in neighboring loci that are so close together that crossing-over never occurs between them. Suppose, by way of example, a mutation took place near the white locus that gave the eye color eosin. If the mutation occurred in a chromosome that had already a white gene, then the new eye color would be due to the combined action of white and eosin. If eosin arose in this way, then, when such an eosin fly is crossed to the original white stock it should give white, since the effect of the new recessive eosin in one chromosome of the F_1 is cancelled by the effect of the normal allelomorph of eosin. In fact, the combination gives an eye color that is not white but is intermediate between white and eosin. If, on the other hand, the mutation to eosin occurred in an X-chromosome not having a white gene, but its normal wild-type allelomorph, then, whenever an eosin fly is crossed to a white fly the female offspring should have red eyes, since each X carries the normal allelomorph of the recessive gene (white or eosin) of the other X. But in fact, an eye-color not red but intermediate between white and eosin results, demonstrating again that eosin and white are not closely linked recessive mutants. The same arguments apply to each of the other nine members of the same series. In the light of these facts it is surprising that statements are still made that the evidence of multiple mutation in the same locus is not established.

Second. There remains another possibility as to the nature of the recessive genes of a multiple allelomorphic series. If the recessive "gene" is due to the loss of material at a given locus then it may be possible that a different amount is lost at each mutation, the amount determining the kind of character produced. It might be assumed that the greater the loss the nearer the approach to white, although this interpretation is not a necessary corollary to the first hypothesis. It might be argued in support of this view that it gives a consistent explanation of the intermediate character shown by the 'compound'. This view meets with none of the difficulties discussed in connection with the first view. It appears to be inconsistent with the general idea of the gene as a particle of definite constitution, suggesting rather that mutations are quantitative. But this can not be used as an objection since the nature of the gene is unknown. On the other hand the facts known in connection with multiple allelomorphs may be equally well interpreted on the view that the gene is itself a unit (rather than an extension of material) that may mutate in several specific directions changing thereby its effects more or less according to the kind of change. Such mutational changes might, it is true, still be thought of as losses (or even gains) to the gene itself in a chemical sense, as by the loss of side-chains etc.

There have been several cases in which a recessive has mutated in a reverse direction either to the wild-type form or to still another allelomorph. These restorations are not easy to account for if the mutation to the recessive involved a real loss rather than a chemical change to another form that is also highly stable but reversible.

The interpretation of dominant characters has also been brought into the discussion. The work on Drosophila has a direct bearing on some of the problems raised since more dominants have been studied in this type than in any other. On the presence and absence scheme, the wild-type gene has been regarded as some material actually present in the germ-cells. Its presence there explains its dominance over — nothing. On the other hand, the appearance of dominant mutations raises theoretical difficulties for the view under discussion. It has been suggested that a new presence, a new gene in fact, arises when a dominant appears. If so, it has no mate, or at least only an absence in the series of existing genes — a hole, as it were, in the alignment would appear to be regarded as its mate. It is not easy to disprove such an interpretation. In contrast to this explanation of dominance the view that seems to have at least the merits of simplicity is that a gene in the wild type may change in such a way that its effects are dominant. The unchanged wild-type gene from which it arose would remain as its allelomorph.

The most serious objection to the above interpretation of dominance is the false emphasis it places on dominance and recessiveness. In fact, there is seldom such a sharp distinction as is implied. While it is true the seven pairs of contrasted characters selected by MENDEL seemed to show complete dominance of one member, the evidence from later work has made plain that this sharp distinction is exceptional rather than the rule; for, in a very large number of cases, perhaps in the majority of cases, the heterozygote is at least slightly intermediate, and accordingly the F_2 ratio is 1 : 2 : 1 rather than 3 : 1. In fact, whether a character is called dominant or recessive is often only a matter of convenience or of convention. When the separation in F_2 between the heterozygotes and one parental type is easier to make than is the separation between the heterozygote and the other parental type, it is customary to call that type the recessive and, hence, call the other parental type the dominant.

CHAPTER VI

MODIFYING FACTORS AND SELECTION

The work on Drosophila has led us to the conception of the individual as resulting from the interaction of a great number of genes — estimated by MORGAN ('22*b*) as certainly more than 2000. As has often been pointed out, any given character must be supposed to be affected by many of these; and in the cases of eye-color, bristle-number, venation, body-color, etc., many separate mutant genes have been discovered and studied that produce specific effects on these characters. Those affecting eye-color were studied in detail by MORGAN and BRIDGES ('13), and those concerned with body-color by MORGAN ('12*d*) — though both of these studies are now very incomplete because of the large number of mutant types more recently discovered. It is a useful conception that each and every character of the organism is due to the interaction of this numerous series of genes, and of the environmental influences to which the individual has been subjected.

For work with linkage, mutation, non-disjunction, and similar problems, it is necessary to select types that overlap the wild-type as little as possible. It thus happens that most of the mutant characters that have been studied and described in detail are ones that are not seriously affected by environmental causes nor by accessory genes. Furthermore, the mutant types commonly used show rather extreme differences from the wild-type, since such types are more easily discoverable in the first instance and are more easily separable in ordinary experiments. It will be shown in the present section that such mutant types are not the only ones occurring in Drosophila, and furthermore, that other, less extreme and less clear-cut, types can be thoroughly studied if special methods are used.

Among the easily separable and slightly variable types it often happens that mutant genes in separate loci affect the same character; there are, for example, eye-color mutations in about 15 loci that are constantly used for linkage work. Flies homozygous for two or more of these mutant eye-colors show a wide range of summation effects. Eosin vermilion is much paler than either simple mutant type, vermilion scarlet is scarcely, if at all, distinguishable from vermilion or from scarlet. It is never safe to predict the color of an untried double mutant type, though some general rules do appear to hold. It has been found that the double mutant types are consistently as light in color as the lighter single type — even such colors as prune and sepia, that are darker than the wild type, have not so far been found to make any other mutant type darker. It is also often true that closely similar eye-colors fail to show a summation effect — e. g., vermilion and scarlet, ruby and garnet.

BRIDGES ('19*c*) has studied a number of mutant genes that produce no appreciable effect except in flies that already have eosin eyes, and others that produce proportionally much more effect on eosin than on wild-type eyes. This case furnishes an unusually clear demonstration of the existence of "specific modifiers", or genes that do not produce an appreciable effect except in the presence of certain other genes.

Several other examples of such specific modifiers have been studied in Drosophila. Perhaps the simplest case is that of the "semi-forked" studied by LANCEFIELD ('18*b*), who found a third-chromosome recessive gene that makes females that are heterozygous for the sex-linked recessive forked, appear intermediate between forked and wild-type. No visible effect was observable in flies homozygous for forked, and only rarely was a slight modification caused by this modifier observed in flies that did not carry the forked gene at all.

CLAUSEN and COLLINS ('22) have studied the case of ski wings (Fig. 16). This character depends for its appearance on the simultaneous action of a dominant gene in chromosome II and a recessive in chromosome III. Neither of these genes produces any visible effect in the absence of the other; together they regularly produce the ski-wing character. The second-chromosome dominant intensifies the ski character when it is homozygous, but is sufficient in heterozygous condition to produce a definite and easily separable character.

BRIDGES and MOHR ('19) have analyzed the vortex character, (Fig. 17) which is somewhat more complex than those so far discussed. There is a second-chromosome recessive gene that was found to be essential for the appearance of the vortex character — but in no case did this gene, voII, produce the vortex character except when reinforced by other genes not usually present in the species. The vortex stock used had a recessive modifier in chromosome III that was usually sufficient, with voII, to produce the vortex character. Sex modified the result, in that females regularly showed the vortex character more strongly than males. Another modifier, probably in chromosome II, was found that intensified the character, and made it appear in flies that

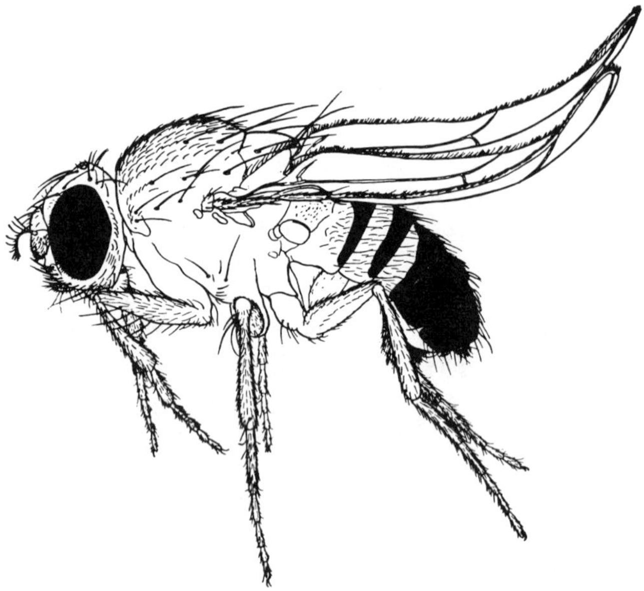

FIG. 16. Ski wings, the "simplex" form, homozygous for a third-chromosome recessive and heterozygous for a second-chromosome dominant.

would not otherwise show it. The dominant mutant gene, streak, was found to act as an inhibitor of the vortex character, but a special modifier was found that largely counteracted the inhibitory effect of streak.

In the cases already cited the modifying genes have been relatively few in number and striking in effect, so that it has been possible to study them individually and locate them accurately in the chromosome maps, by the use of methods only slightly different from those used for genes that do not require any special genetic background before they produce classifiable somatic effects. But there are a number of more indefinite cases, in which more elaborate methods of study are required, since either environmental effects are more easily confused with genetic ones, or more modifying genes are concerned, or the character differences concerned are slighter and less easily classifiable. Special methods for studying these cases have been developed, especially by MULLER('18) and STURTEVANT ('18*a*). These methods depend on determining the linkage of the modifying genes to other, more clear-cut and well-understood genes. By the use of such "standard" genes it is possible to compare sister individuals, which differ genetically in known ways,

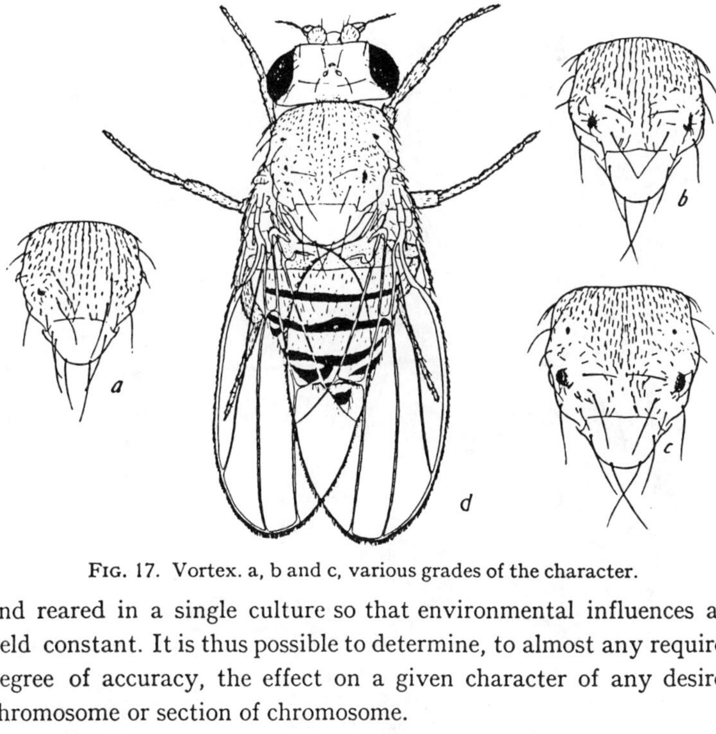

FIG. 17. Vortex. a, b and c, various grades of the character.

and reared in a single culture so that environmental influences are held constant. It is thus possible to determine, to almost any required degree of accuracy, the effect on a given character of any desired chromosome or section of chromosome.

This method has most often been used to study the effect of selection. MORGAN ('19*a*) selected notch, a deficiency producing a dominant character. The notching in the wings, characteristic of the type (Fig. 18), was practically eliminated by selection. Special tests showed that this result was due to one or more third-chromosome recessive modifiers. In the course of the experiments an extreme "short" type of notch appeared, which was shown to be due to a modifier in the X-chromosome. LANCEFIELD ('22) has also reported a specific modifier of the notch character in Drosophila obscura. The modifier, called "split", was a dominant in chromosome III, and produced no visible effect in not-notch flies.

One of the early mutant types found in melanogaster was beaded (Fig. 19). This type was extremely irregular in behavior, and was subjected to selection for extreme beading for a number of years. The resulting stock bred practically true for the beaded character, but still gave anomalous results on crossing. This stock was studied by DEXTER ('14), who showed that the character was influenced by various environmental agents, and also that at least two genes were concerned—one in the second chromosome and one in the third. MULLER ('18) made a much fuller analysis. He showed that the primary factor essential for the beaded character is a dominant in Chromosome III — Bd. This gene has a recessive lethal effect, but the stock bred true because it also had a lethal in its other third chromosome. The balanced lethal system, first thoroughly studied in this analysis, is discussed elsewhere in the present work. MULLER's study also showed that the gene in chromosome II found by DEXTER was a specific intensifier of beaded, and that several other stocks also contained beaded modifiers — some intensifiers and some suppressors, some in chromosome II and some in chromosome III.

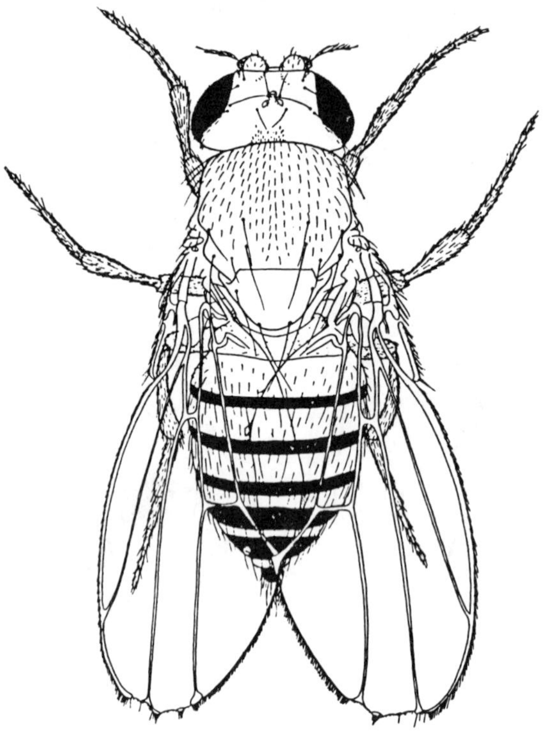

FIG. 18. A slight Notch that shows the Notching in the tips of the wings and the heavier third and fifth longitudinal veins, but does not show the slight disturbances of the bristles, hairs, eyes etc.

A similar case is that of truncate, another of the mutant types discovered early in the Drosophila work. This character, like beaded, was irregular in appearance and was subjected to long continued selection,

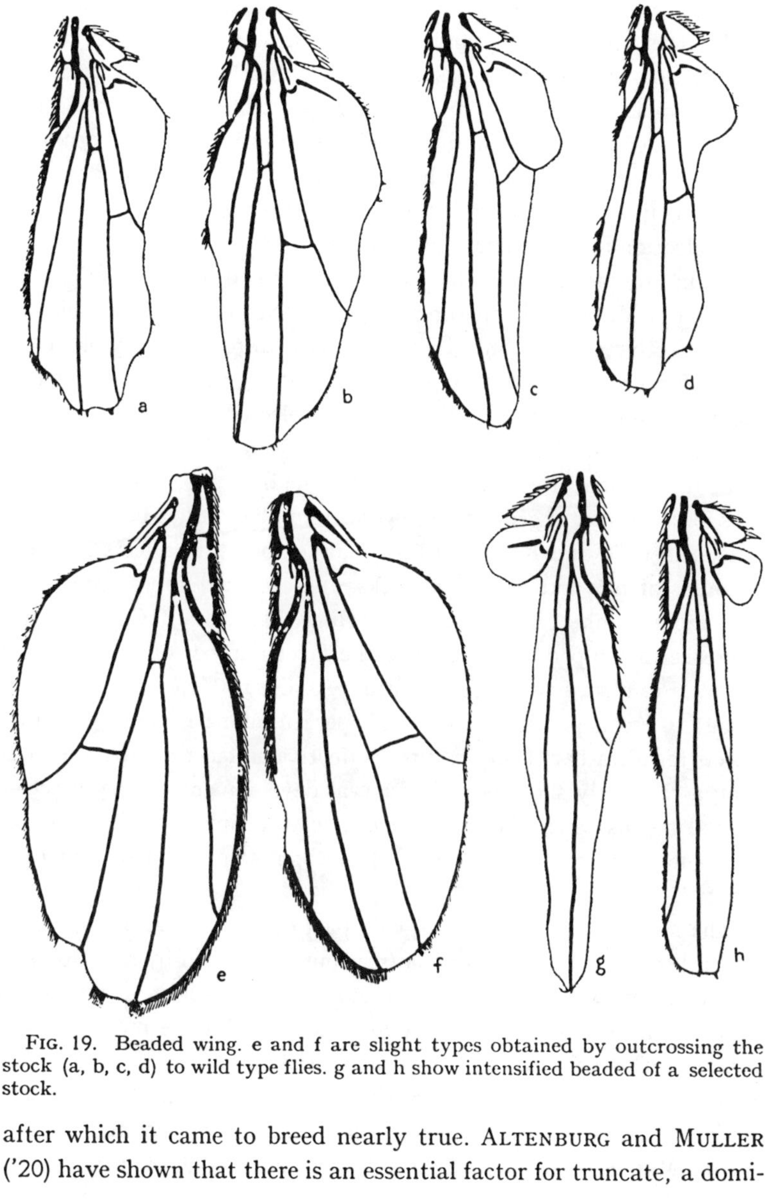

FIG. 19. Beaded wing. e and f are slight types obtained by outcrossing the stock (a, b, c, d) to wild type flies. g and h show intensified beaded of a selected stock.

after which it came to breed nearly true. ALTENBURG and MULLER ('20) have shown that there is an essential factor for truncate, a dominant in chromosome II that is lethal when homozygous. In the truncate stock this gene was balanced by another lethal in the other second chromosome. The truncate character was also intensified in this selected stock by the presence of specific modifiers in chromosomes I and III. Other stocks were found that contained specific modifiers, both intensifiers and decreasers, as in the case of beaded.

The variability in facet number in bar eyed stocks has been intensively studied by ZELENY and his students. ZELENY and MATTOON ('15), MAY ('17) and ZELENY ('22) have shown that selection is effective in isolating lines with high and with low facet number. ZELENY continued the selection for 42 generations of brother-sister matings. In both lines he found mutations of the bar gene to round and to ultra-bar (see discussion of Mutation in the present paper). Neglecting these, selection was effective at first, but gradually ceased to produce any change in mean facet number. ZELENY points out that the mean was changed by elimination of extremes in the direction selected against, not by pushing the extreme further in the direction selected. The results indicate that the original stock, though pure for the bar gene, was heterogeneous for a number of genes affecting facet number. It has been found by SEYSTER ('19) and KRAFKA ('20*a*) that temperature affects the facet number of bar and ultra-bar stocks. Increase in temperature decreases facet-number. KRAFKA's results show that near 17° C., one degree increase in temperature decreases the facet number of bar by about 20, and of ultra-bar by about 5. From 20° C. to 30° the reduction is by about 10 facets per degree rise in temperature for bar, and by about 2 facets for ultra-bar. Temperature is effective on facet number only during a relatively short period of larval development — the effective period begins when about 32 per cent of the time from hatching to emergence has passed, and ends when about 45 per cent of that time has passed.

The number of bristles on a given part of the body is only slightly variable in wild-type Drosophila (see counts reported by MACDOWELL '15, PAYNE '18, STURTEVANT '18*a* and '21*b*); but slight variations do occasionally occur in practically any stock, and some mutant types are regularly characterized by variable bristle number. Since this character can be rapidly and accurately determined on a strict numerical basis, it has been used by several investigators in selection experiments. MACDOWELL ('15, '17, '20) studied the number of dorsocentral bristles in a race derived from wild stock. The normal number of these bristles is four; 48 generations of selection, with brother-sister matings, brought the mean to from 8 to 9. This result was practically gained in the first 8 generations of selection; after that time selection was not successful in altering the mean, in either the plus or the minus direction. After the 48th generation MACDOWELL reared 4 more generations on a large scale without selection, and determined the parent-offspring correlation for the last two of these generations. These correlations were negative in sign, but quite insignificant in value. The crosses made by MACDOWELL show that there was a single primary factor, an autosomal recessive, responsible for the extra bristles. Modifiers were not specifically located, but the data leave no doubt that several were present.

PAYNE ('18) studied the number of scutellar bristles, also normally four. The mean was raised to 9 after 38 generations of selection, the effect being prolonged through about 18 generations, with another increase from the 25th to the 29th. PAYNE's environmental control was not very accurate, and his experiments show that temperature does affect bristle number; furthermore he did not always use brother-sister matings. These two facts may account in part for the long-continued effects of selection, but it seems probable that new mutations affecting bristle-number also appeared in his experiments. Linkage tests showed the existence of extra-bristle genes in chromosomes I and III. In the 6th generation of this experiment occurred a mutation that reduced the number of scutellar bristles. The new type, called "reduced" by PAYNE, is evidently an allelomorph of "scute", a sex-linked recessive. PAYNE ('20) also selected this mutant stock, both for increased and for decreased numbers of scutellar bristles. Sixty generations of plus selection brought the mean to about 3.5; sixty-five generations of minus selection brought it to less than 0.01. In both cases the change was practically complete after 18 generations, though there may have been a further change in the plus line about the 45th generation. Linkage tests showed the plus line carried not only the scute gene, but also plus modifiers in chromosomes I, II, and III. The minus line did not differ markedly from the not-scute stocks chosen for the linkage tests, though there is a suggestion that a modifier was present in chromosome III. PAYNE also showed that low temperature produces a slight increase in bristle-number in scute flies.

STURTEVANT ('18*a*) studied the bristles (dorsocentrals and scutellars taken together) of dichaete, a third-chromosome dominant causing a reduction in bristle number (Fig. 20, *a*). Selection was effective in isolating several plus and minus lines; and, in general, selection in either direction was effective only in early generations of inbreeding. Linkage tests showed that the various lines tested differed in modifying genes located in chromosomes II and III.

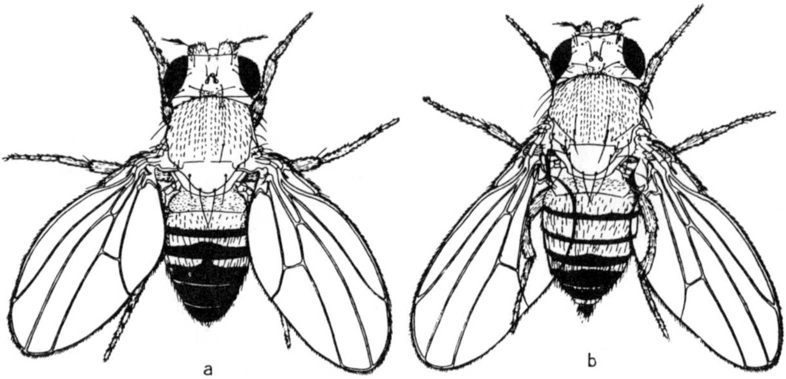

FIG. 20. Dichaete (a) and its allelomorph Extended (b). Dichaete lacks certain bristles from the thorax.

STURTEVANT ('20*d*, '21*a*) reported that hybrids from melanogaster ♀ by simulans ♂ often have less than the normal number of bristles.

Unpublished data show that the number of dorsocentrals or of scutellars in these hybrids is largely determined by the stock from which the melanogaster mother came, though all the parent stocks used in the study have 4 dorsocentrals and 4 scutellars, with only the usual slight variations from these numbers. That is, there are specific modifiers for hybrid-bristle-number. It appears also that there is no correlation between these modifiers and those for dichaete. Of five stocks tested, yellow white happened to give the highest average number of bristles in the F_1 hybrid when crossed to simulans, and the lowest average number in the F_1 dichaetes when crossed to a certain dichaete stock; black vestigial gave the lowest value but one when crossed to simulans, and the highest when crossed to the same dichaete stock. In accordance with this result is the fact that the pattern of bristle reduction is very different in the hybrids from that in dichaete; incomplete evidence also suggests that bristle number is very differently affected by environmental influences in the two types.

LUTZ ('11) carried out extensive selection experiments on slight abnormalities in venation. Lines with a high percentage of extra veins, and others with decreased venation, were quickly established. Reverse selection was also effective even after continued close inbreeding had already occurred. These results leave little doubt that modifying genes were concerned; but the case cannot now be analyzed, as the methods necessary for such analysis had not been developed when LUTZ's experiments were completed.

The male of melanogaster or of simulans has a comb-like row of short stiff bristles on the basal tarsal joint of each front leg. The variability in number of "teeth" in this sex-comb has been studied by CASTLE, CARPENTER, CLARK, MAST and BARROWS ('06) and by HOGE ('15). The number varies from 7 to 14 in individual sex-combs, the average being between 9.4 and 11.1 in all the wild stocks examined. CARPENTER and BARROWS showed that the number of teeth is dependent in part on culture conditions, being greater in the larger flies produced by favorable conditions. Measurements of the length of the right hind tibia were taken as an index of the size of the fly. In four series it was found that this measurement showed a positive correlation with number of teeth in the sex-comb, the correlation coefficients ranging from +.14 to +.71. Number of teeth in right sex-comb was also found to be correlated with number in left comb, the three coefficients determined being +.02, +.30, and +.47. These results show clearly that a large part of the variability in number of teeth is due to environmental causes, and even that much of it is due to causes operating differently on the two legs of the same individual. That there are genetic factors concerned in the variability was shown by HOGE, who isolated both "plus" and "minus" lines by selection. From a wild stock with an average of 10 teeth per sex-comb she obtained one line with an average of about 11.5 teeth, and another with an average of less than 9.5 teeth. No further analysis of the genetic factors concerned is on record.

WARREN ('24) has studied the variations in egg-length found in various stocks. The range in length observed was from. 0.40 mm. to 0.56 mm.; yet even with such minute differences WARREN was able to demonstrate the existence of special modifiers in each of the four chromosomes, though some of the resulting differences in length were only 0.01 mm. A general survey of numerous stocks showed that there is no "normal" egg-length; the two wild stocks reported were near the upper and the lower limits, respectively, of the series. The data also show that in general the egg-length is affected by special modifiers rather than by the mutant genes producing ordinary characters in the imagines. The data of MOHR ('22*a*) show, however, that the gene causing "singed" bristles also causes the singed females to lay short deformed eggs.

The inheritance of fertility was the first genetic problem for which Drosophila was used (CASTLE et al. '06), and since that time much work has been done on the problem by MOENKHAUS ('11), HYDE ('14, '17) and WENTWORTH ('13). All these investigators have studied fertility especially in relation to inbreeding. Various measures of fertility have been used — frequency of sterile matings, total number of offspring produced, and percentage of eggs laid that give rise to adults. MOENKHAUS and HYDE have emphasized that complete sterility is in general quite distinct from low productivity of individual pairs, since the two show no correlation in incidence or inheritance in stocks that carry no known mutant genes. Their results show that inbreeding is apt to be accompanied by an increased number of sterile individuals, but that the sterility can usually be practically eliminated by selection, even though close inbreeding be continued. This result evidently means that most races are impure for certain genes that are capable of producing sterility in some combinations, but that selection and inbreeding may make them uniform for enough "favorable" genes so that sterility is no longer brought about by any recombination effect.

All the investigators mentioned above are agreed that low productivity may occur in inbred strains but it is not a necessary result of inbreeding. It is clear that inbreeding reduces the variability in productivity, but that the average productivity of a given inbred line may be high, medium, or low. If selection is practiced, it is easy to establish highly productive inbred lines. HYDE ('17, Anat. Rec. 6; 506) states that in freshly taken wild stocks from 80 to 100 % of the eggs laid, produce adults, and that inbreeding is regularly followed by a decrease in this percentage. He also reports that when two unrelated inbred stocks are crossed the percentage rises above that of either parent race; while when related inbred stocks are crossed the percentage commonly rises only to about that of the higher parent. This result appears inconsistent with the statement above that it is easy to select a highly productive inbred line. It should, however, be pointed out that "highly productive" is here used to describe a line in which a given pair gives a large progeny. Clearly many factors enter into this result besides percentage of eggs laid that give rise to adults—namely, any factors that influence number of eggs laid. Also, productivity in this sense has not been measured as accurately as HYDE has measured the more specific variable that he uses. In general, while in no case has a complete genetic analysis been carried out, it is clear that wild races are regularly heterogeneous for numerous genes affecting fertility, and that these can be sorted out by inbreeding and selection.

PEARL and others have published a series of studies of the genetic and environmental factors influencing duration of adult life. LOEB and NORTHROP ('16, '17) showed that there is a marked effect of temperature on length of life, such that increasing the temperature between the extremes of 10° C. and 30° C. causes a marked decrease in length of adult life. PEARL and PARKER ('22) found that as many as four successive anesthetizations did not appreciably decrease length of life. The same authors ('22) report a slight increase in length of life due to increased ventilation, but no effect from embryonic juices either of the chick or of Drosophila larvae. PEARL, PARKER, and GONZALEZ ('23) found that clipping off the wings somewhat decreased the duration of life. PEARL and PARKER ('23) report that there is an optimum "density of population" for length of life, such that in a one-ounce vial about 35 flies will live longer than will a number somewhat smaller (2 to 20) or larger (65 to 200).

HYDE ('13) showed that length of life depends in part on inheritance, and presented some data on the result of crossing different strains. His results cannot be fully analyzed. PEARL and PARKER ('22) found that offspring of separate pairs from a given stock might show significant differences in average length of life; and when the descendants of the separate pairs were bred as separate unselected stocks, later tests showed that the initial differences between the lines persisted. Evidently the original stocks were heterogenous for modifiers affecting length of life.

PEARL, PARKER, and GONZALEZ ('23) and GONZALEZ ('23) crossed a long-lived wild stock to a short-lived multiple recessive stock—black purple vestigial arc speck. Various combinations of F_1 and P_1 individuals were mated, and duration of life was determined for the resulting offspring. In general it is clear that most of the difference between the two parent stocks (with respect to length of life) was due to the vestigial gene and its normal allelomorph. There is also evidence that other genetic differences were present.

Frequency of crossing over in a given chromosome section has been shown by GOWEN ('19) to be highly variable. The coefficient of variability is unusually high in comparison with that found for other biological phenomena. Numerous studies have shown that this variability is in part due to environmental causes, in part to genetic ones. It was shown by BRIDGES ('15) that crossing over in the middle portion of chromosome II is less frequent in old females than in young ones. This result was verified by PLOUGH ('17), who also showed that crossing over in the same region is increased by temperatures either above or below the usual range of room temperatures. As shown by PLOUGH ('21) and by unpublished data, both the age and temperature changes are specific, occurring only in certain regions of certain chromosomes. Another environmental effect has been reported by MAVOR ('23) and by MAVOR and SWENSON ('23), due to X-rays. The treatment given decreased crossing over in chromosome I, increased it in chromosome II.

Genetic differences in crossover values have been recorded by STURTEVANT ('19*a*), MULLER ('17, '18), BRIDGES and MORGAN ('23), DETLEFSEN ('20), DETLEFSEN and CLEMENTE ('23), WARD ('23), GOWEN and GOWEN ('22), PAYNE ('18), and others. These results make it clear that striking differences in cross-over values are sometimes brought about by genetic causes, and also that relatively slight differences are sometimes inherited. The bearing of these facts on theories of crossing over will be discussed elsewhere in this paper (Chapter X).

Several mutant characters besides those already discussed have been shown to be modifiable by environmental conditions. Some of these have also been shown to be modifiable by selection, and it seems likely that in general a character subject to variability will be easily influenced by either type of agent (genetic or environmental), as has been pointed out by MULLER ('18). A partial list of such variable characters that have been recorded follows:

Vestigial. CASTLE ('15) cited experiments by WENRICH showing that the variability in wing-length of a vestigial stock was increased by crossing to wild and extracting. ROBERTS ('18) failed to modify the character by selection, but found that high temperature increased the wing-length. The latter result has been verified in unpublished experiments carried on in the Columbia laboratory.

Abnormal abdomen. MORGAN ('15*c*) showed that this character is most extreme in flies reared under moist conditions, and that it practically disappears in flies reared in dry conditions.

Reduplicated legs. HOGE '(15) found that this character was much more marked, and appeared in a much higher percentage of the flies, if the cultures were kept at a low temperature. The data indicate that the low temperature was effective chiefly during early larval life.

Bent (melanogaster and virilis). METZ ('23) exposed cultures to low temperature (9°—12° C.), and found that certain characters (rough eyes, abnormal bristles) were accentuated in both species. The effective period was apparently late in larval life or else just after pupation. Unpublished results also show that the bent character in melanonasgter is subject to variability due to genetic causes.

Abdominal spotting (busckii). WARREN ('20) found that the spots on the abdomen of busckii could be increased or decreased in number by selection, and that temperature also affected their development.

Variable eye (hydei). HYDE ('22) found that this character was extremely variable in appearance, and was most extreme in flies reared under warm and dry conditions.

MULLER ('16) kept a number of recessive genes in heterozygous condition for many generations. One of these, dachs, was then recovered from the stock, and the extracted homozygous dachs flies were compared with flies from a stock that had been kept homozygous for dachs. It was found that the extracted flies showed the dachs character in a slightly more extreme form than did those from the homozygous dachs stock. This result was not conclusive, because based on only a few flies; but MARSHALL and MULLER ('17) obtained the same result on a larger scale with balloon and with curved, extracted from the same heterozygous stock. They concluded that in the homozygous stocks natural selection operates on modifying genes in such a manner as to shift the mean grade of the mutant character toward the more viable wild type. It seems likely that this often happens in stocks of variable mutant characters. In at least three cases — those of bent and of eyeless in melanogaster and of vesiculated in simulans — it has been observed that unselected homozygous stocks gradually became nearly like the wild-type in appearance. But in all three cases, when these stocks were outcrossed to other stocks not carrying the mutant gene in question, F_2 showed a high proportion of individuals with extreme forms of the mutant character.

CHAPTER VII

VIABILITY AND LETHALS

One of the most conspicuous features of many mutant types of Drosophila is their weakness. This weakness may express itself in various

FIG. 21. Rotated abdomen. The abdomens of both sexes twisted to the left through 60 to 90 degrees; both sexes unable to mate.

ways, such as slowness of movement, lack of response to ordinary stimuli, likelihood to get stuck in the food or in the moisture in the bottle, or, in general, in shortness of life. At present we can only ascribe these changes to effects of the mutant gene upon internal structures over and beyond the particular visible somatic effects produced.

There are certain types of characters shown by mutant races of Drosophila suggesting that the further a character departs from the normal the less viable it is. Thus, some of the eye-colors that depart only a little from the color of the wild-type (vermilion, for example) do not appear to lower the vitality of the race, while a greater change (white-eye, for example) is associated with a weaker type. On the other hand, such a principle does not always apply. The vestigial race is as viable as certain other types of wing modifications (truncate or rudimentary, for example) where the wing is less obviously altered. Nevertheless, in a general way, experience has shown that in Drosophila a very slight modification seems to affect the general viability of the mutant less than an extreme one, and this applies to those visible structural changes that do not, as such, put the individual at a disadvantage on account of the nature of its modification.

In a few stocks semi-lethal types appear, i. e. types that only rarely emerge from the pupa stage. These are interesting because, when fertile, they make possible the study of the influence of two semi-lethal genes in the same individual. Lethal-10 (Bridges '17*a*) and lethal-20 and dwarf-warped are such semi-lethals. Lethal-11 and lemon are not so extreme.

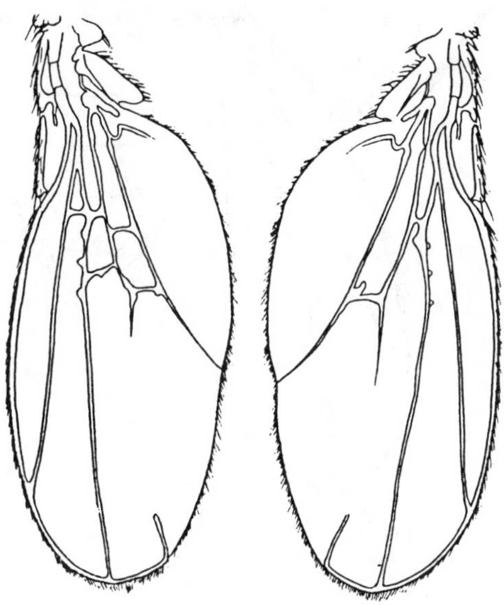

Fig. 22. Cleft venation, showing projections from third longitudinal vein beyond anterior crossvein, and break in fourth longitudinal.

Mohr ('23*a*) has described a sex-linked mutant that he calls chlorotic (which he locates at 0.1) whose viability is so low that he places it amongst the lethals or semi-lethals. The surviving males (and females also) are fertile, and give a 1 : 1 ratio. Since the chlorotic males have a low viability, the sex-ratio is affected in proportion to the number that survive. Similar effects upon sex-ratios by low viability of mutant sex-linked types have been described by MORGAN ('12*e*, rudimentary), by MORGAN and BRIDGES ('16 lemon), by BRIDGES ('16, spoon and '17*a*, lethal-10).

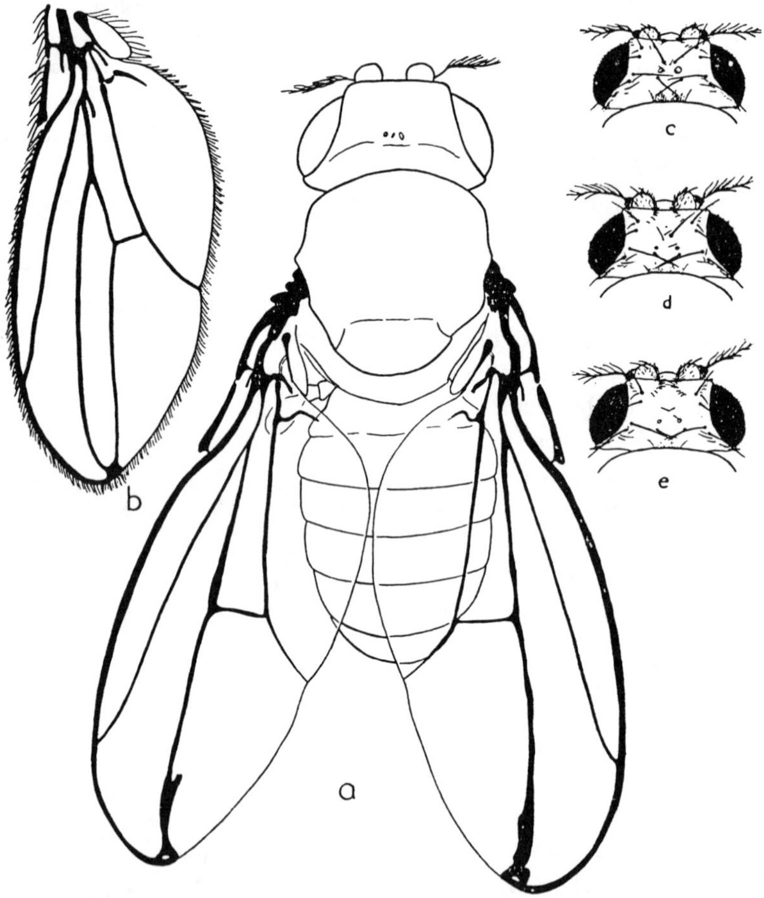

FIG. 23. Fused venation, average-type of wing at b, extreme at a. Different grades of reduction in ocelli and ocellar bristles at c, d, and e.

Certain stocks are partially or wholly sterile, and the sterility may occur in the female or in the male. A few examples of each kind are as follows. The females are entirely sterile in dwarf (Fig. 25), morula, singed, ovaless, reduced, and in the suppressor of Hairy-wing. The males are sterile in rose, parted, rotated-penis, cleft (Fig. 22), and in giant (BRIDGES, MORGAN, LYNCH). In some cases the sterility is due to a structural defect in the ducts or gentalia (Fig. 21), but in others, few or no eggs mature. Lozenge is largely female-sterile. In some of these female-sterile cases there are allelomorphic types that are female-fertile, such as scraggly-bristles, (allelomorphic to the female-sterile reduced-bristles), singed ^{2} and singed ^{3}, and lozenge ^{4}. There is also one fairly fertile allelomorph of rudimentary (r^{12}). Most "Minutes" are rarely fertile in the female, but fertile as males, as in Minute-f, Minute-g, Minute-l, and Minute-m. This also holds for fused (Fig. 23), rudimentary (Fig. 24), divergent (Fig. 25) and no-wings (Fig. 26).

There is one important consideration connected with the interpretation of the nature of the sterility and viability effects that, if misunderstood, may lead to serious consequences as to Mendelian work in general. It is this. Any alteration in the composition of the germ-material, whether the effect of mutation in a single gene or of a change in a number of genes (by loss or by increase), that brings about alterations

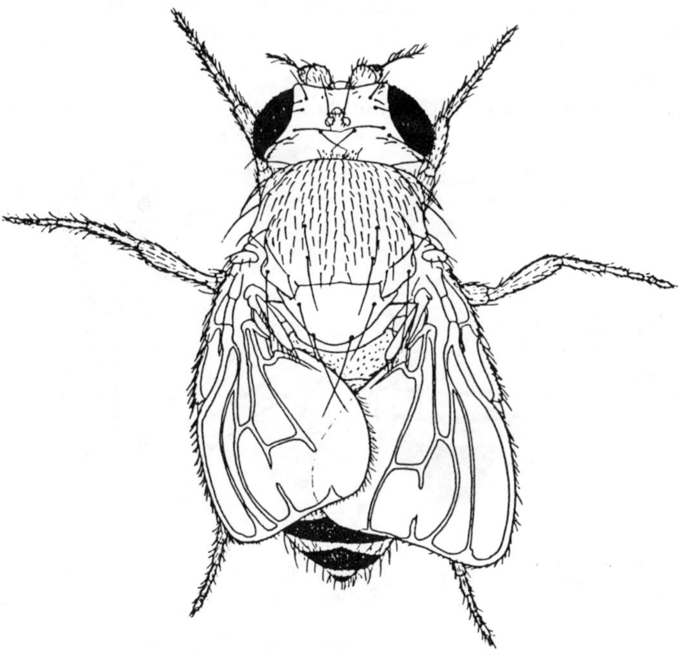

FIG. 24. Rudimentary.

in organs whose normal functioning is of vital importance to the individual, will probably lead to the death of all the individuals or of many of the individuals whether the change occurs in the embryonic or in the adult stage. Between the extreme changes resulting in death of all individuals, and those that weaken but do not destroy, gradations are expected and are in fact found in the Drosophila stocks. At one extreme are those cases when the embryo dies before hatching, then those that die as larvae, then those that die in the pupa stage, then those that emerge only rarely (semi-lethals) and the graded series of mutant stocks some of which live only a few days, others that live longer but are relatively weak in constitution, and lastly those that can scarcely be said to be weaker than normal flies. We speak of the inviability of those stocks in which the number of resulting individuals is below standard, limiting the use of the word lethal to cases where the adult stages are not reached because of death in the egg, larva, or pupal stages.

When one recalls that in the normal development a great many genes

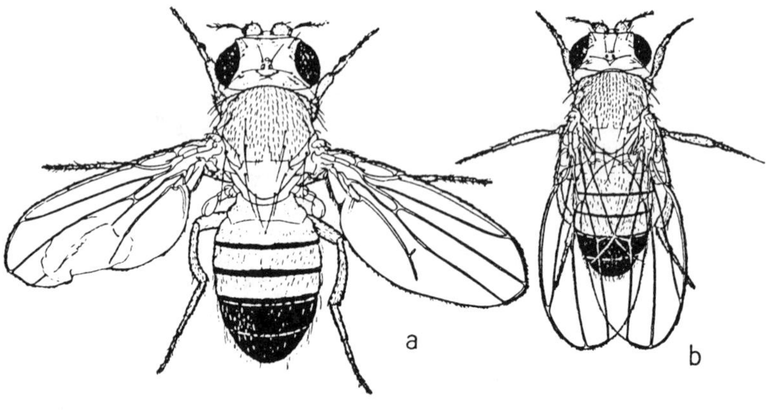

FIG. 25. Divergent wings (a) is ♀-sterile. The dwarf race (b), drawn to same scale, is also ♀-sterile.

must be supposed to be involved in the formation of each organ or part of the body, and when it is also recalled that the organism is an extremely delicately balanced affair, passing through an intricate series of interdependent stages in its embryo and pupa, the expectation would be that most mutations would probably be lethal in effect. The evidence that such may be the case will be discussed later.

As yet only a few lethals have been studied with respect to the stage at which the lethal effects injure or destroy the individual. Enough is known, however, to indicate that some lethals bring about their effects in the pupal, others in the larval stages and others in the embryo before it hatches from the egg. It is possible that in other cases, still earlier stages of development are affected.

In contrast to these lethal effects stand out all the other Mendelian characters that have survived to maturity because they effect only slight changes in the individual, or affect such parts as are not essential to the life of the individual under domestication. Many of these mutant races would die out if left to themselves under natural conditions. Two corollaries follow from these considerations. First, it is erroneous to conclude that mutation involves only superficial characters (or trivial differences). It is true that these are the ones that are most worked with in plants and animals. They are used because of their survival value, not because they differ in kind from other mutations that involve essential characters. Second, it would also be erroneous to conclude, because most changes in the germ material produce deleterious or injurious results, that all mutations must be of this kind. It can be demonstrated that wild stocks often carry different factors that modify one or another part, and that are inherited as any other Mendelian character. Therefore, since no other kinds of characters are known that

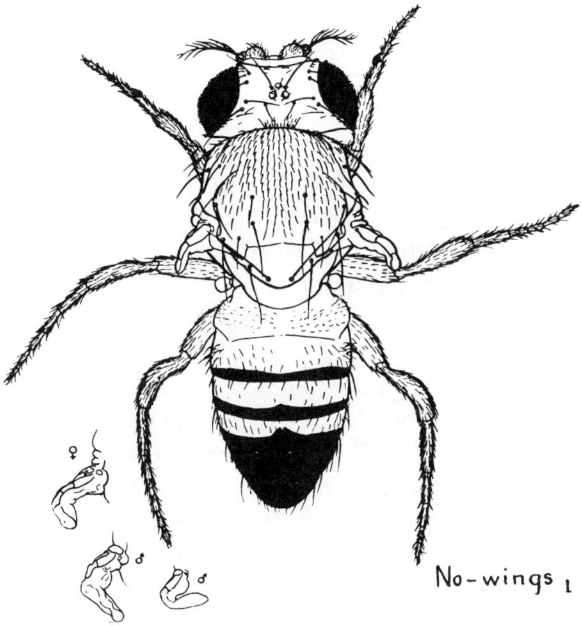

FIG. 26. No-wings, showing type of wings, of balancer and of scutellar bristles.

have a different type of inheritance, it is highly probable that all characters have arisen through gene mutation with the exception of those due to alteration in the ratio of genes.

Sex-linked Lethals

To certain kinds of genes the term lethal has been applied. These genes are supposed to destroy the individual, not necessarily because the lethal gene makes a poisonous substance, but more likely because it brings about some change, either in the germ-cells or in the embryo, that interferes with normal development. When the effect is on the germ-cell it is spoken of as a gametic lethal. When on the embryo (i. e. on any stage after fertilization) it is spoken of as a zygotic lethal. No gametic lethals are known in Drosophila, but several kinds of zygotic lethals have been recorded. The simplest cases are recessive lethals in the X-chromosome, and these were in fact the first recorded. Here a change has taken place in a normal gene of such a kind that any male (the male has but one X-chromosome) dies if the gene is present. In other cases the lethal effect may be due to the absence of a whole chromosome, or to the occurrence of extra chromosomes; to the absence of a part of a chromosome, or to the presence of an extra section of chromosome. The effect of these and of other kinds of lethals can best be described in connection with an account of their occurrence.

The earliest work on Drosophila in which a change in the sex-ratio was observed that depended on the presence of a lethal was that of RAWLS ('12). The demonstration that these unusual sex-ratios were due to a gene in the sex-chromosome was made by MORGAN ('12) with RAWLS' material. RAWLS found one female that gave 222 daughters to 110 sons, and another female that gave 308 daughters to 131 sons. Of the daughters 63 were mated separately to their brothers. Twenty-six pairs had high ratios and thirty-seven had normal ratios. This equality of normal and of aberrant cultures, as we now know, is expected on the basis of a sex-linked lethal. The ratios found by RAWLS that were still higher than 2 : 1 were probably due to the presence in the mother of two lethals, one in each of her X-chromosomes. The few males that came through we now explain as due to crossing-over.

The demonstration that the deficiency in males is to be accounted for by the hypothesis of a lethal factor was as follows (MORGAN '12).

Red-eyed females of RAWLS' stock were mated in pairs to white-eyed males. Some of the pairs gave a 2 : 1 ratio. When their daughters were in turn bred to white-eyed males half of them gave again a 2 : 1 ratio distributed as follows:

red ♀	red ♂	white ♀	white ♂
448	2	452	374

The deficiency is obviously in the expected red males that should carry the grandmaternal X-chromosome. The two red males can also be accounted for by crossing-over. It was pointed out that if a lethal factor were carried in one X-chromosome of any female that gives a 2 : 1 ratio, the results are accounted for. Further evidence was obtained as follows: All the red-eyed daughters, in the case given above, should contain one red-producing chromosome from the mother, that carries the lethal factor, and all should give a 2 : 1 ratio irrespective of the mate to which they are bred. This was found to hold. Similarly, the white-eyed females should all give a 1 : 1 ratio, which they did. The two crossover males should not transmit the lethal to their descendants if they were crossovers. This also was found to be the case. Experiments with other mutant genes confirmed the evidence that the locus of the lethal was near that of white.

A further experiment gives confirmation of the above interpretation. Some of the females from RAWLS' flies were bred to miniature males. Some of the daughters, from those mothers that gave a 2 : 1 sex ratio, were bred to miniature males again and half of them gave sex-ratios with the following distribution:

Total ♀	Long-winged ♂	Miniature ♂
910	156	243

The most deficient class is that of long-winged males, and this is the expected lethal class. The flies in this class are relatively more numerous than in the last experiment with white eyes because, as was pointed out, there is a *much higher crossover value* for the lethal and miniature. In the former experiment the crossing over with white was only once in 200 times, hence the lethal gene must have been near white, and therefore should give a relatively high crossover value with miniature.

An account of the lethal, called lethal-1, is given in the paper on Sex-Linked Inheritance, by MORGAN and BRIDGES ('16). Its location was put at 0.7. In the same paper the inferred occurrence of a second lethal (lethal-1a) in RAWLS' very high ratios was discussed. This second lethal must have been in the homologous chromosome, at level very near to that of the first one, since only about 2% of males came through. This places the gene about 2 points from the other lethal. If to this evidence is added the evidence from the number of daughters that RAWLS found carried or did not carry one of the lethals, the location is estimated at 3.3, to the right of lethal-1.

The interpretation of the occurrence of two lethals in the same female (both in the same or in opposite X-chromosomes) was first discussed in a second paper by MORGAN ('14). It was there pointed out that a second lethal might arise by mutation in the germ track of a female already possessing one, or be derived from the sperm-cell of the father of such a female, where it may have arisen by mutation in the testis. It could not have been present from the beginning in all the cells of such a male, for the male himself would not have survived. A third sex-linked lethal (lethal-3) was reported (MORGAN '14) to be located at 26.5 and a fourth lethal at 19.5 (See MORGAN and BRIDGES ('20, page 74, 75). The occurrence of still another lethal, exceedingly close to the white locus, seems probable from data obtained at the same time (its occurrence is discussed by MORGAN and BRIDGES '16, page 76).

In order to discover whether lethal factors are more frequently found in laboratory stocks that have been inbred for several or many generations in mass cultures Miss STARK ('15) undertook to examine the frequency of lethals in such stocks as contrasted with wild stocks. No lethals were found in the wild stocks examined, but four were present in the inbred stocks. A discussion of the origin of these lethals and their location is given by MORGAN and BRIDGES ('16, pages 64—66). One is placed at 16.7 and another at 23.7 (STARK '16).

Another sex-linked lethal (lethal-4) was reported by BRIDGES ('16, page 149) at 49. He also reported at the same time the occurrence of a lethal (lethal-7) that produces black masses in those male larvae that carry this lethal. Larvae that develop these black bodies, later called tumors, die before the pupa stage. The location of the gene was not reported at that time farther than to place it to the left of eosin. Later work locates the gene of lethal 7 at 0.1 (BRIDGES, STARK '18, page 287). This lethal stock was turned over to Miss STARK to study the origin and development of the tumor, its structure and its effect, if any, on the larvae. The results of this work (STARK '18) showed that the black bodies owe their color to the development of pigment in embryonic or imaginal cells. These cells multiply rapidly before the time of the appearance of the pigment, as well as after that event. The pigment is largely confined to the surface cells that form a laminated membrane over the mass of interior polygonal cells. Owing to the movements of the larvae, the older tumors may be carried to different regions of the body. The death of the larvae was supposed by STARK to depend upon the state of development of the tumor, but not on the number or the size of the tumors. On the other hand death may be due to the condition of the larva, the tumor being only an index of the lethal effect of the tumor itself. Miss STARK tried to determine whether the tumor itself kills the larvae by some product or by-product. Attempts to inject the tumor material into normal larvae proved difficult. The mortality of the larvae is so great when the body wall is punctured, even by minute glass tubes or by very small hypodermic needles, that the results are somewhat uncertain. The black bodies were removed from many larvae (about 1600) by making a minute cut in the body wall. Only 5% of the larvae survived the operation for 24 hours. None of the survivors pupated. As a control the same sort of operation was made on nearly as many normal larvae (1400). Here also only about 5% survived, but those that lived pupated and formed normal flies. The result may mean either that the tumor had already weakened the larvae, or that they will die regardless of the tumor, which would then be interpreted as only one of the symptoms of the lethal gene — an indicator of its presence. Miss STARK interprets the result as showing that the removal of the tumor had a beneficial effect on the surviving larvae, since unoperated larvae seldom lived as long as those of the same size that were operated upon, but as no exact data are given on this point the conclusion may be questioned.

Attempts were also made to insert the tumors removed from l_7 larvae into normal larvae. As a control, small bits of sterilized charcoal were injected into normal larvae. Five per cent of the controls lived and pupated. Likewise 5 % of those with inserted tumors survived the operation, but died before pupation, which Miss STARK suggests is due to the inserted tumor. This conclusion is, however, doubtful since slight differences in the ease of injecting the charcoal as compared with the insertion of the tumor might account for the death of the latter, or possibly the body fluids that went in also with the tumor might have acted injuriously. That such differences may occur is suggested by a further experiment of Miss STARK. Tumors were washed in LOCKE'S solution and a small amount of the mixture was forced into the body cavity of normal larvae by means of a capillary pipette. Out of 100 larvae, 40 survived the operation. No visible tumors developed, but all died within 12 hours. On the other hand when an equal number of normal larvae had LOCKE'S solution injected by the same method, 50 % survived and emerged as normal flies. Whether the difference in the two results is due to fragments of injected tumor or to the body cavity fluids of the l_7 larvae, that may be injurious, is not evident from this test. In the light of other evidence to the effect that the size of the tumors in the l_7 larvae bears no obvious relation to the death of the larvae it may appear improbable that the results are due to the tumor substance itself.

In order to meet the possible criticism that the tumors are due to infection Miss STARK isolated eggs whose surface had been sterilized. These gave the same proportion of tumor larvae as the controls, not sterilized. The fact that only males, and only half the males, die is itself evidence that death is not due to infection. Furthermore the evidence that only half the daughters and only those that can be shown by linkage tests to carry a particular chromosome, transmit the tumor to half their sons is further evidence that the results are due to inheritance and not to infection.

Many experiments made by Miss STARK to destroy the tumors or to make them innocuous by subjecting the larvae to X-rays gave negative results.

When the lethal-7 stock was examined several years after it had been used by Miss STARK, larvae with tumors were absent, yet from the classes of individuals present and from the sex-ratio it was evident that the gene was still there. The l_7 gene had been run against yellow and eosin, between which the lethal locus lies, i.e. a female had been made up with l_7 in one X, and yellow and eosin in the other X. Her gametes carry one or the other X. Mated to a yellow eosin male she produces two kinds of daughters, one like the mother that has red eyes and the other yellow eosin, as follows:

$$\frac{l_7}{y\ w^e}\ \text{♀ by}\ \frac{\qquad}{y\ w^e}\ \text{♂}$$

$$\frac{l_7}{y\ w^e}\ \text{♀ and}\ \frac{y\ w^e}{y\ w^e}\ \text{♀ and}\ \frac{\qquad}{y\ w^e}\ \text{♂}$$

Only the red-eyed daughters carry the lethal-bearing chromosomes. If these are again bred to the only kind of sons that come through, viz. yellow eosin, in successive generations, the l_7 can be carried in the female line, and has been maintained for several years by this kind of selection. In each generation the l_7 sons are represented by larvae that die. The three loci, y, l_7 and w^e are so near together that even single crossing-over is extremely rare, and is easily detected when it occurs.

Now when this stock was later examined (as stated above) no "tumor" larvae were found, and since the class of males that carried that gene was also absent, it follows that the l_7 gene was still producing its effect.

The tumor can, however, be brought out at any time by crossing the female heterozygous for l_7 to males of other stocks. Some of the male larvae then show tumors. From the outcrossed flies, stock can be secured that shows the tumor again. This result appears to mean that some other factor, or factors, have appeared in the l_7 stock and cause the l_7 larvae to die in the egg or soon after hatching. While this result does not show that it is not the tumors in the larvae that kill them, it shows at least that the l_7 gene in conjunction with modifying factors can kill without the production of a visible tumor.

Muller and Altenburg ('19) carried out experiments that were planned to test the frequency of the occurrence of lethal factors. A number of females (90) were tested and the offspring of those that carried lethals were discarded. Daughters of the females that gave normal sex-ratios were again tested. These were also chosen from as many separate families as possible. Thirteen out of 385 females showed a sex-linked lethal. Since some of the ratios on which the conclusion rested were doubtful (despite the fact that they were tested) a better type of experiment was carried out. Heterozygous females $\frac{WVF}{w^e vf}$ were bred to wvf males (the preliminary test of the mother having been made). If either the WVF or the w^evf chromosome contained a lethal (that came from either parent) then the corresponding class of sons would be killed. Here both the 2 : 1 ratio test as well as a class test is possible. Out of 1082 "families" tested, twenty showed lethals; a ratio of 1:53. The flies that gave these results had been kept at two different temperatures, one at about 66° F. and the other about 80° F. The former gave one lethal to 90, the other 1 to 40. Two other lethals came from flies at room temperatures. On the assumption that lethals occur as frequently as this in the other chromosomes, the authors calculate that about "one fly in every thirteen has a new lethal mutation in some chromosome or other." MULLER again returned to the discussion of this topic in 1920, '23 and '24.

The earlier work in Drosophila reported above had shown that lethal mutations in the X-chromosome are frequently met with. The linkage test, whenever made, indicates that the lethals are different. Only when they have nearly the same position in the chromosome it is difficult to prove their independence. Their identity can not be proven, when they are identical, by the method used for ordinary cases of suspected new characters, viz. by breeding them to each other, because sex-linked lethals kill all males.

A sex-linked lethal that kills the female when homozygous has been described by D. H. THOMPSON ('21). Its presence in the male can be recognised by its effects on the wings and legs. The wings are held erect and the second pair of legs is feeble. When a heterozygous female is crossed to an "erect" male the offspring are in the ratio of 1 female to 2 males. The locus of this lethal is at 38. Furthermore, THOMPSON found a modifying factor that makes the erect dominant in heterozygous females.

BONNIER ('22) located two lethals (l_1 and l_r) in the sex chromosome at 7.5 and 16.4 respectively. He gives formulae by which when two lethals are present they may be located, but states that the more satisfactory way to locate them is to separate them and locate them singly in the usual way. A third lethal (l_p) is placed at 15.7. In a later paper BONNIER ('23*a*) by combining the l_1 and l_r lethals in different ways and with other strains produced different sex-ratios. A new sex-linked lethal (l_g) that kills only females is described and approximately located at 63.1. The origin of this lethal has some interest since it was detected in the offspring of mothers that carried l_1 and l_r in the same X-chromosome and might be expected therefore to give a 2 : 1 ratio. But two of these mothers gave a 1 : 1 ratio. From the constitution of the flies it seemed certain that a lethal had arisen that was sex-linked and killed the females. This was confirmed by tests upon the heterozygous offspring.

Lethals that are not Sex-Linked

Lethals in other chromosomes than the X-chromosome are more easily overlooked, since they are discoverable only when they happen to be near to a gene that is under observation, and an F_2 culture is being raised. In the paper on the second-chromosome group by BRIDGES and MORGAN ('19) four lethals are discussed. The truncate studied by ALTENBURG and MULLER arose from beaded stock. Because of the small number of normal flies thrown by highly selected truncate stock, it was inferred that a lethal was present in the not-truncate chromosome-II. In other words, the truncate had become a balanced stock.

Another second-chromosome lethal, called dachs-lethal, was found in experiments dealing with Star and dachs. Certain cultures in which one-third of the offspring were expected to be dachs gave no dachs. It was pointed out, that the lethal effect may also be accounted for on the view that a deficiency had occurred in the dachs locus (see BRIDGES and MORGAN pp. 277—283). Still another lethal (lethal-IIa), located supposedly between black and curved; was discovered. (BRIDGES & MORGAN pp. 286—91). Subsequent work has shown that the locus is far in the right end, and that the former location was wrong because of a linkage variation. A peculiarity of the situation when a recessive and a lethal are in opposite chromosomes is described in this paper, namely, that the percentage of recessives in F_2 is a constant value, 33.3, irrespective of how much or how little crossing-over there is between the loci.

In the third-chromosome paper of BRIDGES and MORGAN ('23) several autosomal lethals are described. It is pointed out that certain earlier results with pink gave aberrant ratios that were probably due to two lethals rather than to poor viability of pink.

One of these, lIIIa, is associated with CIII, and this combination of a lethal and a crossing-over suppressor has frequent use in artificially balancing such stocks as Stubble, Delta, Hairless and Beaded that are lethal when homozygous, or indeed any poorly viable or somewhat sterile recessive mutation in the right half of chromosome III.

The combination lCL serves a like purpose for the left half of chromosome III.

The application of these views concerning lethals to the case of beaded wings gave a satisfactory solution of a case that had been difficult to harmonize with Mendelian formulae.

The beaded stock arose in 1910 (MORGAN '11) but did not breed true to the character. Mated to normal sisters a beaded male produced 1.6% beaded. When inbred they gave 3 % beaded offspring, and again inbred, 8.5 % beaded. Selection of beaded was continued for many generations. Finally a beaded stock was obtained that gave nearly 100 per cent beaded. The stock was turned over to DEXTER ('14) for further study. He found that the chief factor for beaded is in chromosome III (near the right end). An intensifier was found in chromosome II. DEXTER did not find that flies pure for beaded died, i. e. the main gene in double dose is lethal. Environmental factors were also found, that affect the development of the character.

MULLER ('17, '18) made an exhaustive study of beaded and worked out the essential genetic facts as to its peculiar inheritance. MULLER succeeded in showing that the chief factor for beaded is lethal, killing all flies that are homozygous for the factor; hence the difficulty experienced at first in obtaining a pure beaded stock. The results of selection that led to a stock practically pure for beaded were due to the appearance and isolation of other factors, more especially, of a lethal (lIIIa) in chromosome-III. This lethal kills all flies homozygous for it, but in heterozygous form has no effect on the character beaded. It appeared in that member of chromosome-III that did not carry beaded, and its locus lay about 10 units from the beaded locus. The beaded stock would then be expected to give about 10 % normal flies in each generation, but at the end of the selection period, i. e. at the time when MULLER took up a study of the selected material not even ten percent of the offspring of beaded flies were normal. This is due, as experiments showed, to the presence in the stock of another mutant factor called CIII, which when heterozygous almost entirely prevents crossingover in the right limb of the chromosome. Moreover, this suppression is most effective in the region where the lIIIa and CIII lie. Since CIII lies in the same chromosome as the lethal, it remains in a heterozygous state. Owing to this peculiar genetic situation both types of homozygotes, namely beaded wings and lIIIa, die. This is due to the action of lethal factors lying in opposite chromosomes viz., beaded and lIIa. Such a condition has been called one of balanced lethals. Since its discovery the principle involved has been extensively used in Drosophila stocks to maintain certain types that could be kept otherwise only by selection in each generation.

MULLER also pointed out how a process of continued selection of a dominant like beaded, that is lethal when homozygous, would be likely to lead to a condition of balanced lethals. Most of the dominant mutations in Drosophila are lethal when homozygous. MULLER points out that this is probably not a peculiarity of dominant genes as contrasted with recessive genes, but because lethal mutations are themselves more common than other kinds, (for a discussion of this question see page 23) and when dominant, may manifest themselves in the heterozygote in some visible character. If in such a stock that is being selected a recessive lethal appears in the chromosome that is homologous to the one carrying the first gene, i. e., in the "opposite chromosome" it will lead to the appearance of a smaller number of individuals selected against, hence such lines will be chosen as parents of the stock. Thus, almost automatically a selected line becomes a balanced lethal line if nothing else stands in the way of its formation. How pure the line will become will depend on the distance apart of the two lethal loci. Crossing-over also breaks the apparent but superficial purity of such a line. Such occurrences might have been put down as due to atavism before we came to understand what atavism may mean. If there appeared in nature a balanced lethal stock that rarely gave a cross-over type, the apparently new type might be called a new mutant. In fact it seems not improbable, as MULLER pointed out, that some of de VRIES' mutant types of Oenothera were mutant types released by crossing-over.

Balanced lethal flies crossed to other races will give, in the first generation, two types of hybrids. Thus, beaded crossed to wild type gives 50 % beaded, and 50 % normal. When one balanced stock is crossed to another balanced stock twin hybrids are produced that may or may not be constant. Ever sporting stocks will arise in balanced stocks if a mutant appears or is introduced in one or the other of the chromosomes carrying a lethal. The genes fail to manifest themselves owing to the enforced heterozygosis, but if the gene crosses over from the lethal factor so that it may later get into a homozygous combination its effects will, so to speak, come to the surface, and were its origin unknown it would appear to be a new mutation, or if it had been introduced by a carefully planned experiment one would speak of it as having been recovered.

A similar situation was found in another stock called truncate (ALTENBURG and MULLER '20).

When the list of dominant mutations of Drosophila is compared with the list of recessive mutations, the much higher proportion of lethals amongst the dominants is at once apparent. The first impression is that there is something peculiar to dominants that makes them lethal. In fact, this relation had led to an interpretation that put dominants in a class apart from other mutations; but a closer examination of the evidence removes the mystery. In the first place, it is to be remembered that the distinction between a dominant and a recessive factor is largely an arbitrary one, depending to a great extent upon the line of demarkation in the F_2 grouping of the three classes of AA, AA', and A'A'. In the second place, MULLER has pointed out that lethal mutations are more numerous than any other kind of mutations, but very few of them are ever seen as mutant characters since they are usually discovered only by changes in the ratios of the classes. On the other hand a lethal mutation that affects some visible part of the body when in a heterozygous condition is readily observed and kept as a mutant, and only later, as a rule, is it discovered that the homozygous condition is lethal. In other words since the known dominant lethals affect some visible character they are discovered in far greater numbers than are recessive lethals which are usually overlooked. Since we have no reliable census of the frequency of the two kinds of lethals there are no compelling arguments in favor of the view that a lethal effect is more often associated with dominants than with recessives. The fact that some dominant characters, are not lethal is also to be kept in mind. The following dominants belong to this group. In the first chromosome Abnormal abdomen, Beadex, Bar, and Hairy-wing; in the second chromosome Cream-II, Ski-II, Lobe, Lobe^2 and partly lethal Streak and Curly; in the third-chromosome all dominants are lethals when homozygous except Deformed, which rarely survives.

CHAPTER VIII

ABNORMALITIES THAT ARE NOT INHERITED

Many exceptional, abnormal or monstrous types of Drosophila have been met with in our work, and while some of them have been recorded many are here described for the first time. Most of these have been met with more than once, and some of them occur frequently. There is no certain rule by which such modifications can be distinguished from mutations. Breeding tests are essential to settle this question, but in general when an abnormality appears only on one side of the body it is probable that it will not be inherited; yet there are exceptions to this, and asymmetry is not always a safe guide.

A very high percentage of these aberrant types is found persistently in certain mutant stocks, and may in part be looked upon here as one of the by-products of the mutant gene. It is impossible, therefore, to completely exclude many of these abnormalities from the heritable characters; but whenever they can not be made to reappear in considerable numbers of the offspring by breeding from parents that show such modifications, or have not been produced by special environmental conditions, they may be looked upon as accidents of development. It is rational to suppose that certain conditions of development may increase the chance of such accidents happening more often in certain mutant types than in other types. Expressed in another way, it may be said that the modifications characteristic of certain types make it more possible for the machinery of development to run off the beaten track.

Under these conditions no attempt is made in what follows to sharply distinguish between abnormalties that are not inherited and those that recur rather frequently as by-products of certain mutational changes. For purposes of easy identification, the material is arranged according to the organs in which the abnormalities occur. It would perhaps be better to attempt to arrange these aberrations in other ways, such, for instance, as their association with certain mutants or

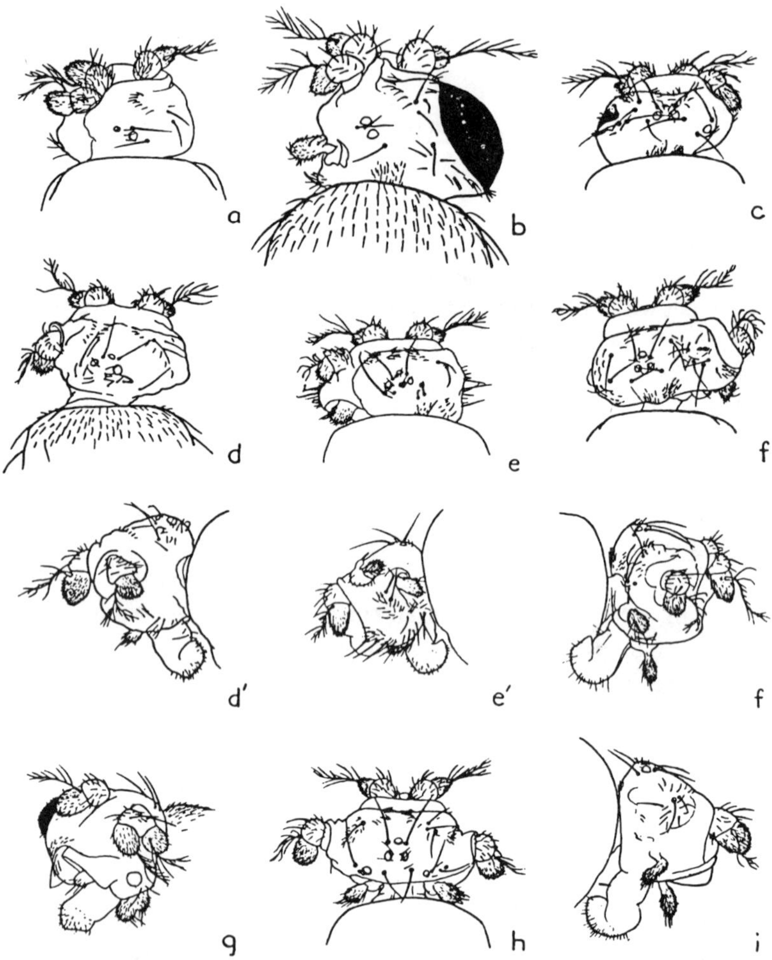

FIG. 27. Antennae doubled (a), or replacing eyes (h). Extra palpie in many Figs. c is from Bar stock, b is from wild-stock, and the remainder from eyeless.

with their frequency of recurrence, etc, but for our present purposes this does not seem necessary. It would also be quite difficult to carry this plan through systematically.

Antennae. Duplications of one, or even of both antennae sometimes occur, almost exclusively in eyeless stock. (Fig. 27). The terminal portions only, may be doubled. In these cases the duplicated part is a mirror figure of the antenna of the same side. Rarely an additional antenna

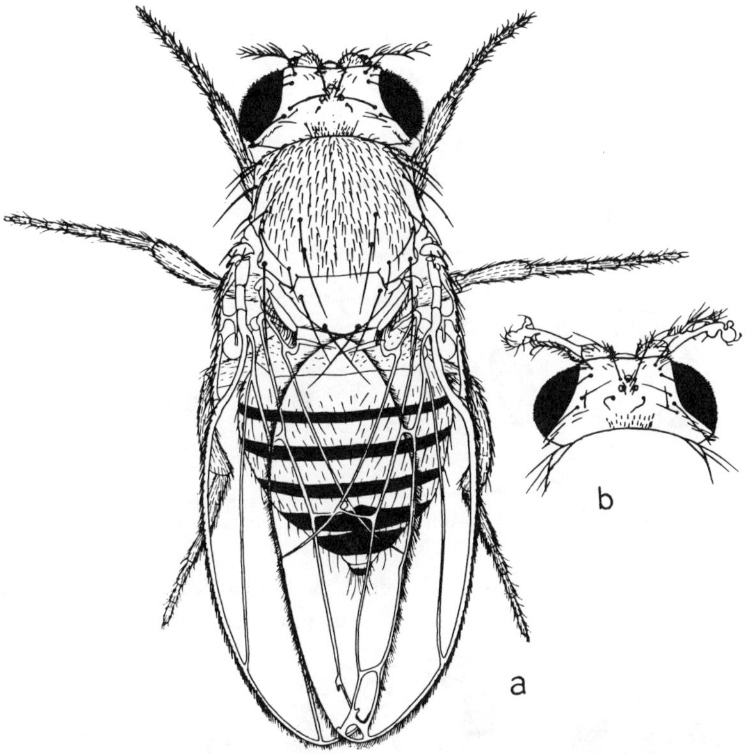

FIG. 28. Pointed-wing. a. change in venation at tip o ɟwing, which is pointed; b, occasional outgrowth of antennae into wing-like rudiments.

appears in place of an eye. It shows a serial homology with the antenna of its side i. e. the two are not mirror figures. These additional antennae are frequent in the eyeless stock. Very rarely two additional antennae are present on the same side in place of the eye and these are mirror figures of each other. Not infrequently when both eyes are absent an additional antenna is found on each side. There is another frequent modification associated with these additional antennae of eyeless. At the lower edge of the empty eye-space, a palpus is present (Fig. 27*b*, *e*, *e′*, *f*, *f′*, *g*, *h*, *i*,) and furthermore at the top of the head, just posterior to the empty eye-space one or more ocelli are present, often on a stalk

(Fig. 27, *h* and Fig. 32). These relations suggest, in fact, that the eye-space represents imperfect duplication of the head. Whether all these parts develop out of the imaginal disc or discs that give rise in normal flies to the compound eye, or whether contributions are made to the region from the imaginal discs of the palpi and ocelli could only be determined by an embryological study.

Occasionally the ends of the antennae are dark or even black, but experiments have failed to produce from these flies a black-antenna stock. As flies get old the antennae sometimes become dark and sometimes drop off.

The two antennae are fused at the base in many flies of the cut 6

FIG. 29. a and b, normal eyes divided into two parts; c. Bar eye, with extra small eye and ocellus on stalk.

stock. This condition is so frequent as to be a feature of the stock rather than an abnormality that is not inherited. It probably occurs

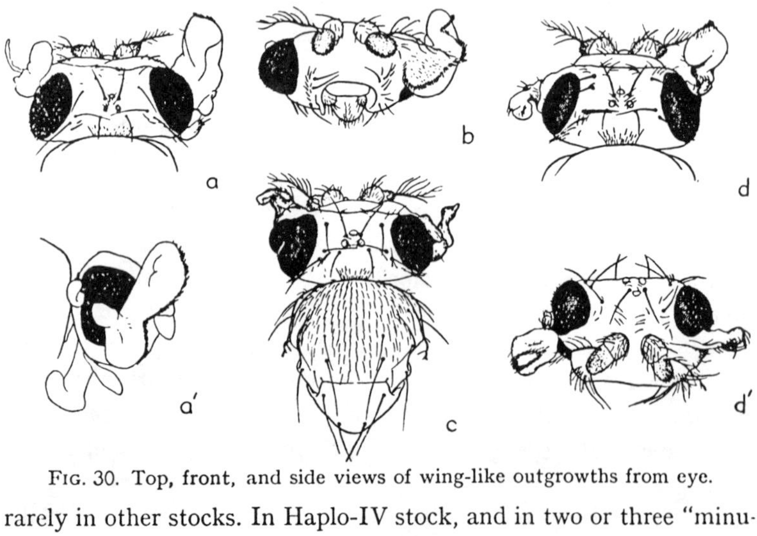

FIG. 30. Top, front, and side views of wing-like outgrowths from eye.

rarely in other stocks. In Haplo-IV stock, and in two or three "minutes" the arista is thickened and jointed and has a row of bristles with a pair of terminal hairs that resemble the claws at the end of the legs.

In a stock called pointed-wing (Fig. 28) the aristae of the antennae are occasionally modified, as shown in Fig. 28*b*. There are some features of this modification that suggest the outer margin and base of the wing, but the evidence is insufficient to establish such a comparison.

Mouthparts. The palpi are sometimes black but attempts to produce a true-breeding race have failed.

Compound Eyes. One or both of these eyes may be reduced in size or absent (Fig. 27*b*). Occasionally one eye may be divided into two parts each completely circular in outline (Fig. 29*a*, *b*), each made up of ommatidia that are normal in size, hence there are fewer of them in the smaller part. In some flies a faint seam may be seen on the surface of the eye due to irregularity in the line of connection of the rows of ommatidia. In one of the eyeless stocks a wing-like organ

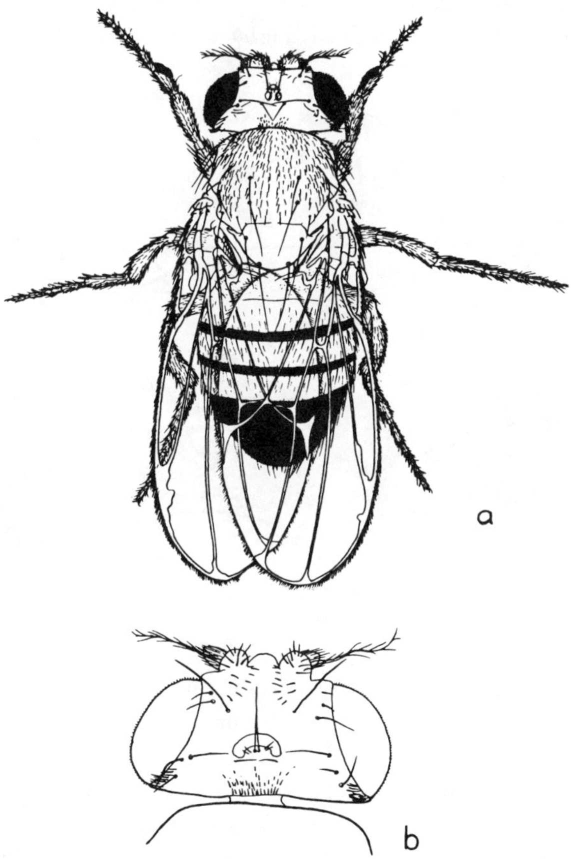

FIG. 31. Delta. a, venation delta-like and hairs on thorax irregular; b, ocelli often run together into crescent which encloses the ocellar bristles.

arises from the middle region of the eye (Fig. 30). It can not be stated with certainty that the protruding organ is an imperfect wing, yet the arrangement of the bristles along its edge and its bag-like transparent material suggest at least such a comparison. The occurrence of antennae in place of the eye in eyeless flies has been mentioned above.

Ocelli. The ocelli are sometimes displaced and one or more may be absent. In "delta" they often run together (Fig. 31, *b*) forming a crescent-like structure. This occurs also in a mimic of delta called deltex. In fused, the ocelli are often abnormally arranged, and one, two, or all three are occasionally absent. In eyeless there are often scattered ocelli on the top and back of the head. Sometimes they have short stalks (Fig. 32).

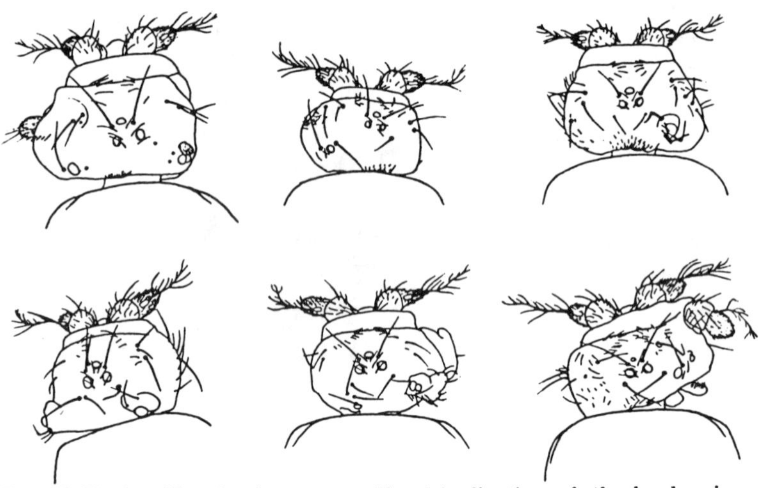

Fig. 32. Eyeless flies showing extra ocelli and duplications of other head-regions.

Thorax. The dorsal portion of one side of the thorax may be absent (Fig. 33). In such cases the rudiment of the wing on the defective side may be seen through the anterior wall of the abdomen (Fig. 33*b* and c) In these flies the head and eyes appear to be larger than the eyes of sisters of the same approximate size (Fig. 33*b* and *a*). These flies with half of the upper thorax absent are not at all infrequent in vestigial stocks, although the deformity is not inherited.

The thorax may be split in the middle line so that its two halves are quite separate.

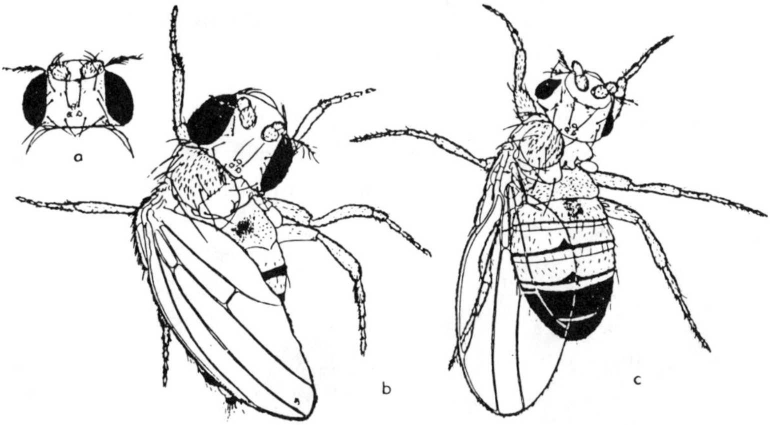

FIG. 33. Half-thorax. b, bent female with right dorsal part of thorax absent. The head in such cases is usually extra large, as seen by comparison with normal sister shown in a. c. eyeless fly, in which dislocated material is in abdomen.

Legs. The tarsus of one or both of the hind legs may be shortened and thickened and blackened. Some obstacle to their extension in the pupa seems to be responsible for this crippling. Rarely one leg may be absent. Duplications of whole legs or terminal parts of legs have been found (Fig. 34a). One fly was found in which the two front legs were united as far down as the tarsus. It did not breed. Fig. 34b, b′.

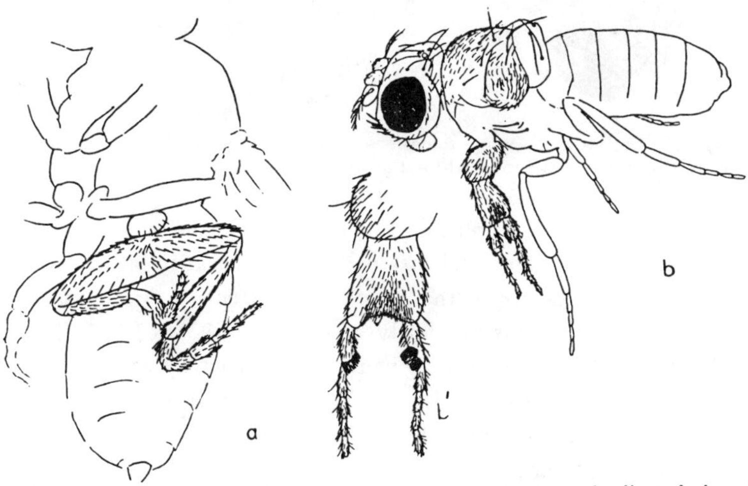

FIG. 34. Reduplicated and fused legs. a, left posterior leg reduplicated; b and b′, fusion of anterior pair of legs with partial reduplication of left; thorax also abnormal and without wings or balancers.

In certain intersexes the bases of the fore-legs may be fused in the middle line, also in pointed-wing.

Wings. When the fly emerges, the wings are folded and lie as two dark pads on each side of the thorax. Occasionally they fail to expand. In one stock "club", this occurs in about 20% of the flies (Fig. 35a).

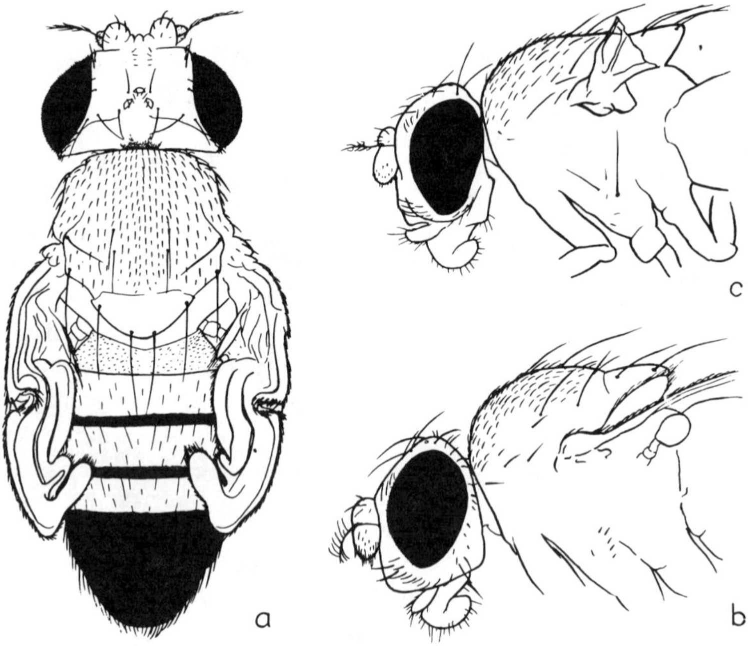

FIG. 35. Club wing. a, wings remain as in pupa in about 20 % of cases; b, in all flies of club stock sternopleural bristles are absent, as seen by comparison with normal fly at c.

One or two cases of duplication of the wing on one side have been observed (Fig. 36). The duplication of both wings as in bithorax stock is an inherited feature of that stock. The extent of development of these posterior wings is extremely variable in bithorax (Fig. 37).

The margin of the wings may be torn in such a way as to resemble beaded. A nick may occur in one or both wings near the end. It is difficult to distinguish between such nicks and others like that of notch which is a characteristic of a stock. Occasionally the wings are bent down in a way to resemble the inherited bent-down wings of arc. Irregularities may be found in the venation, aside from those characteristic of certain stocks. A short branch from the posterior cross vein sometimes occurs. The variation appears so frequently in certain stocks as to represent an inherited character, but in other stocks it does not appear again. Sometimes a newly hatched fly has a drop of fluid in one or both wings, and when the fluid disappears and the upper and lower surfaces come together the wing is wrinkled. This is a characteristic development in balloon and blistered stock. In flies that are over-etherized the wings turn up over the back.

FIG. 36. Reduplicated left wing, parts of three wings being present.

Balancers. These are not often carefully observed. It was at first sup-

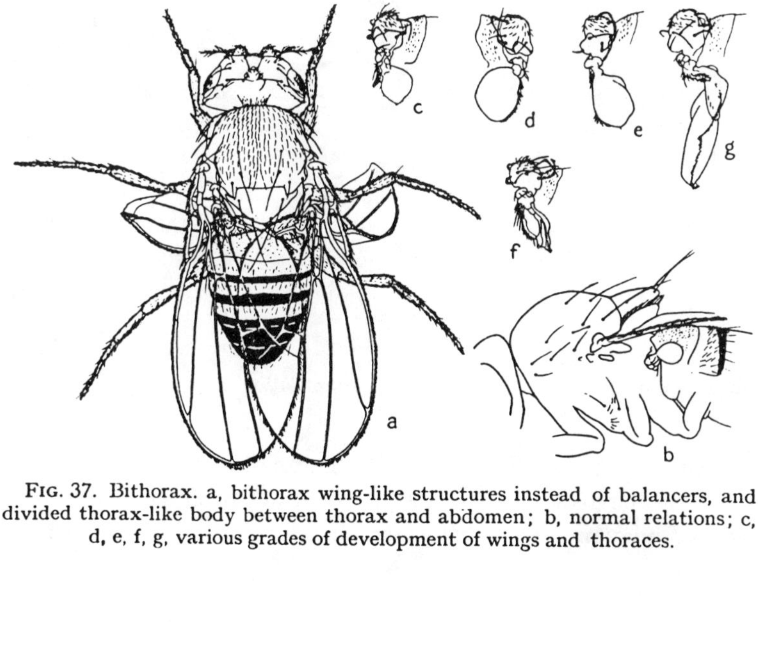

FIG. 37. Bithorax. a, bithorax wing-like structures instead of balancers, and divided thorax-like body between thorax and abdomen; b, normal relations; c, d, e, f, g, various grades of development of wings and thoraces.

posed that they were changed into wings in bithorax (Fig. 37). A more thorough scrutiny showed that bithorax is due to the total replacement of the third part of the thorax by a repetition of the second part that normally carries wings.

In one instance (Fig. 38a, d) a winglike structure developed on one side in place of a balancer. In another case (Fig. 38c) there was a normal balancer on one side and a double balancer on the other side, one of them (the anterior one) being somewhat wing-like. In still another case Fig. 38b the terminal section of one balancer was drawn into a wing-like plate with two veins.

Abdomen. The abdomen, especially in young flies, has a whitish tip and the colored bands are faint. Later, pigment develops. The white tip occurs frequently in certain stocks and may be due to multiple factors. In other stocks the black bands at the end of the abdomen of the female may be so broad that they resemble the male. This also is a feature of certain stocks, and in them may be a genetic character. In other cases it may be only an extension of the variable condition of the pigmentation. The abdominal sclerites may fail to meet above in the midline of the abdomen (Fig. 39a, c) or two consecutive rings of the same side may meet in the mid-dorsal line, or one or more bands may

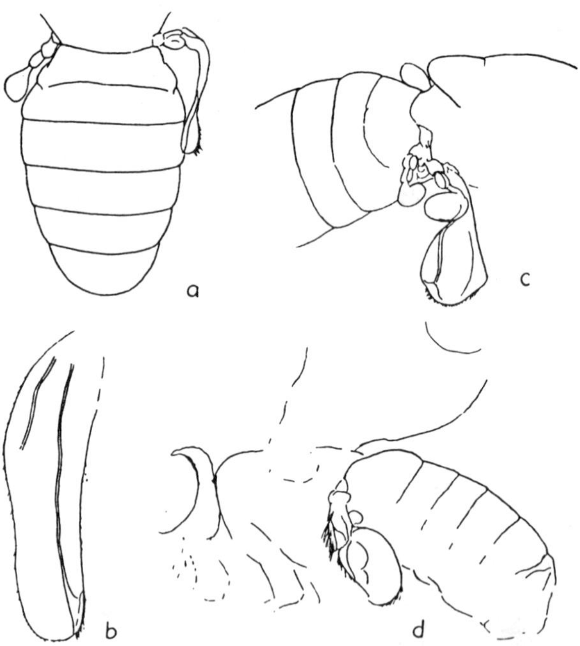

FIG. 38. a, b, d, Balancers wing-like; c, reduplication of section bearing balancer.

be absent on one or on both sides so that the total number of abdominal segments is reduced (Fig. 39b). In one stock (yellow white) in which irregularity of the bands was quite often present it seemed to be caused by high temperature. In another stock this or a similar character appeared to be inherited either as a very variable character or as a multiple factor case (MORGAN '15*c*).

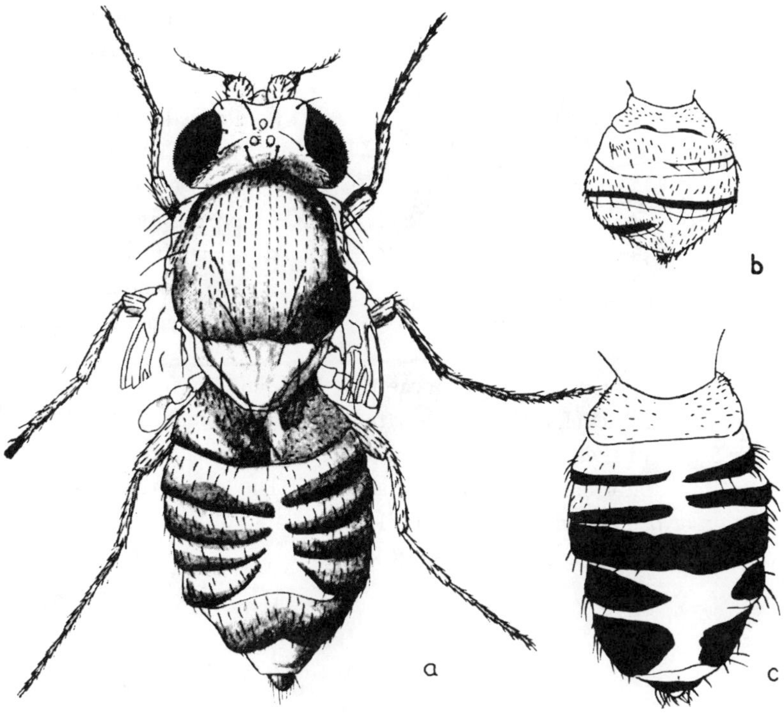

FIG. 39. a and c, incomplete union of parts of segments growing up from the two sides; b, omission of segments.

Genitalia. Occasionally a fly is found that seems to have no genitalia and no anus. When such flies are dissected the genitalia may be found inside the tip of the abdomen. In some way these organs have been turned in or are overgrown in the development. The twisted penis of one stock is described elsewhere. The irregularities shown by intersexes, and by gynandromorphs fall into other subdivisions (Chapters XIV, XV, XIX). Two cases in which the genitalia were reversed are shown in Fig. 40.

Senile Changes. Most of these have been mentioned above. The flies get darker as they get to be old, and the antennae and balancers may fall off. Some of the bristles also may fall off or become broken. The change in the color of the eye of old flies characteristic of certain mutant stocks is referred to elsewhere (Chapter XXII).

Juvenile Characters. The protrusion of the front of the head (ptilinum) that pulsates as the fly emerges from the pupa case may be delayed in its contraction. The wing pads may be delayed in expanding or the abdomen in contracting. The scutellar bristles may remain crossed.

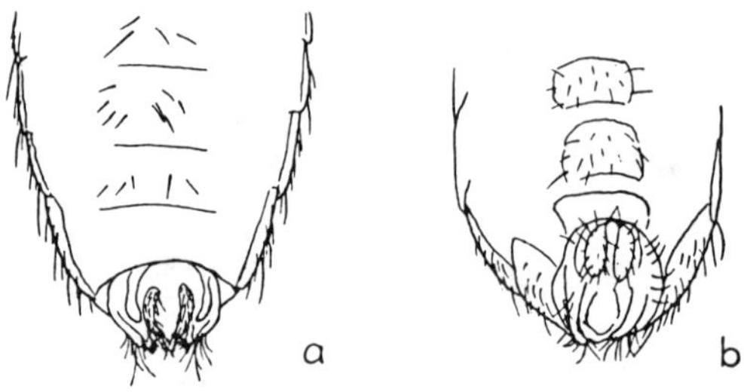

FIG. 40. Genital and anal segements reversed by rotation through 180°.

Larva. Starved larvae, or larvae that have left the food before becoming fully developed, may pupate and produce small flies that lay few eggs, but these normal in size. Larvae have been so little examined that few abnormalities have been recorded. The lethal tumor larvae are described elsewhere (Chapter VII). The cause of high death rate of larvae in crowded conditions is not known.

CHAPTER IX

INFERENCES CONCERNING THE NATURE OF THE GENES

Between the characters shown by the individual and the postulated genes in the germ cells lies the whole field of embryonic development. MENDEL's theory and its later developments are not primarily concerned with this field — i.e. they are not immediately concerned with the way in which the genes bring about the results that we see in the characters that are affected. In fact, embryology has thus far developed in entire independence of the new principles of heredity — principles that are nevertheless the fundamental postulates for any account of the developmental process that pretends to be anything more than descriptive or pictorial. Even although we do not pretend at present to know much concerning this relation between genetics and development, still the genetic work has progressed far enough to indicate some of the important considerations from which the student of embryology must proceed. These may be categorically stated as follows:

1) All the heritable characters about which we have definite information at present are referable to the genes. The only rare exceptions to this rule are the few cases where characters such as chlorophyll grains are transmitted from one cell generation to the next by self perpetuating "plastids". Hence, one of the central problems of embryology will be the discovery of the methods by which the genes effect the development of the characters of the individual.

2) Since the cytological evidence shows that all the chromosomes are transmitted to all the cells of the embryo (except when chromosomal elimination or accidents of distribution occur), it follows that the development of every part of the embryo must take place in the presence of the entire genetic complex.

3) Since those mature eggs of the heterozygote that retain the recessive factor, when fertilized by sperm of a recessive male, give rise

only to the recessive character, it follows that whatever influence the dominant gene may have had on the cytoplasm before it was eliminated in the polar body, is changed before the adult characters of the individual appear. It follows that the differentiation of the cells that determine the characters is traceable to the gene in the chromosome, or, in other words, the cytoplasm is controlled by the genes. Two qualifications of the statement are necessary ((4) and (5)).

4) In the earliest stages of development, such as in the tempo of the cleavage and in the form of the cleavage, there is evidence (rather meagre at present) to show that these stages have been already determined by the cytoplasm of the egg under the influence of the maternal chromosomes. This means that the entering sperm-nucleus has not had sufficient time to affect the cytoplasm of the egg. In later stages of development the influence of the paternal genes becomes evident. These conclusions are important for embryology since they carry with them the refutation of a view that threatened for a time to give an erroneous interpretation of the relation of Mendelian inheritance to development; namely as stated by BOVERI that the genes affect only the more superficial characters of the individual and not the fundamental ones, which reside in the cytoplasm.

5) The above considerations must not be interpreted to mean that the cytoplasm is without significance for the problems of development. On the contrary, since the actual movements of the cytoplasm and of the mitotic changes, as well as differentiation, occur in the cytoplasm the composition of the latter at any moment is the paramount agent that determines the changes taking place, and this is the visible or physical activity that constitutes the very essence of development. It is quite possible that in the cytoplasm there are transmitted self-perpetuating units through which the action of the gene becomes possible. Nevertheless the outcome is, in the final analysis, referable to the composition of the genes. We are at present quite ignorant as to how this relation is estableshed.

Certain more negative conclusions may also be added:

(*a*) Whether the genes are active all the time, or whether their activity is correlated with certain phases of development, we do not know at present and can scarcely even surmise.

(*b*) Whether the genes can influence the cytoplasm during the resting stages of the nucleus or only when, at cell division, they are, for a short time, free in the cytoplasm, is not known.

(*c*) Whether the genes act as enzymes or catalyzers or set free other substances that permeate the cytoplasm and influence its activity is unknown.

(*d*) Whether the gene is a chemical molecule or a collection of molecules is guess work at present. The only attempt to determine its size — a problematical procedure at best — may now be referred to.

The work of crossing-over in Drosophila has offered an opportunity to attempt to calculate limiting values of size for the gene. These calculations are based on data that are not only meagre, but in certain aspects quite inadequate; yet admitting that the procedure is hazardous, it may not be uninstructive as the first attempt to get even an approximate estimate of the size and number of the genes. (MORGAN '22b).

(1) The head of the spermatozoon contains a haploid set of chromosomes. A measurement of its volume would give the maximum possible size for one set of genes. If the total number of genes could be determined this number divided into the volume of the sperm head would give the upper limit of the genes. The volume of the head of the spermatozoon can be approximately determined, but the number of the genes is difficult to calculate. Two methods have been utilised in the attempt to determine their number.

(*A*) Assuming that one gene is as likely to mutate as another, then if we know how many mutations have occurred in loci that have previously mutated, we can calculate how many genes are present. In fact we have kept a record of the number of times the same mutation has recurred, or at least we have a fairly complete record as to the frequency of mutation at a given locus. If this information were complete it would be large enough at present to furnish the needed data, but it is avowedly incomplete in several respects. It is probable that many repeated mutations have been overlooked, and those mutations that produce lethal effects can be only nearly located but not with sufficient accuracy to identify them as the same recurrent mutant type. With these limitations it is estimated (MULLER see MORGAN '22) that there are more than 2000 genes in Drosophila. Since each individual has two sets of genes there will be 4000 in all. This calculation leaves out of account the reverse mutations of the bar gene, and as stated, the lethal mutations. Multiple allelomorphs are counted as independent mutations.

(*B*) A second method of determining the upper limiting size of the genes depends on crossing-over. If two genes are so close together that crossing-over between them is an exceedingly rare event then this crossing-over value expressed as distance may be taken to indicate approximately the nearness of the genes to each other. The drawback here is to determine where the lowest crossing-over limit lies. All degrees of crossing-over occur between genes, and the lowest values are difficult to determine, for one reason, because crossovers are so infrequent that the data may not be significant. Since, however, we find these values gradually approximating to the lowest values, that value taken as the lowest may not be far from the true value. Assuming then that the lowest value is about one-fifth of a unit of the genetic chromosome we can for any given chromosome determine what part of its genetic length this constitutes. For instance, in chromosome II there are 108 units. It must contain therefore about 5 times as many genes or 540 in all. If we calculate the volume of chromosome II from its length and diameter and divide the volume by 500, the diameter of the gene is found to be $\frac{20}{1000}$ of a micron.

Again by dividing the calculated volume of all four chromosomes (0.0236 cubic microns) by the estimated number of the genes (2000), the diameter of the gene turns out to be 60/1000 of a micron.

These calculations make the size of the gene larger than the estimated size of the molecule of haemoglobin, but not much larger. There may be protein molecules larger than those of haemoglobin, and, if so, the genes may be no larger than such molecules, but such comparisons have at present very little value, and little weight is to be attached to them.

CHAPTER X

LINKAGE

Amongst the 400 mutant characters of Drosophila melanogaster about 150 show a type of inheritance different in certain respects from that of the others. Thus, when any one of them that is recessive enters a cross from the male side it reappears in half of the grandsons and in none of the granddaughters. Furthermore, if such a character enters a cross from the maternal side, it appears in F_1 in all of the sons and in none of the daughters. In F_2 it then appears in half of the grandsons and in half of the granddaughters. These characters follow the same type of transmission as does the X-chromosome. These 150 characters constitute a group of "sex-linked" characters, and the genes are assigned to the X-chromosome or chromosome I. When any two of these characters enter a cross in P_1 from the same parent they tend to emerge together in F_2. From this relation they are said to be "linked" to one another.

Linkage is also found between characters that are not sex-linked; and these characters fall into three linkage groups. The "second" group is defined as that in which the characters are linked to the mutant black (the first mutant of this group used in linkage tests) and to each other, but not to any other characters. Likewise the third group includes those that show linkage to pink and to each other, but to no other characters. The fourth linkage-group corresponds to the small fourth chromosome of Drosophila melanogaster, as has been demonstrated by experiments in which one of these chromosomes is absent.

Linkage Groups and Crossing-Over

Owing to the fact that all the known characters of Drosophila melanogaster fall into four linkage groups, the study of the inheritance of linked genes has received far more attention than has the independent inheritance of characters belonging to different groups. If the linked genes never separated from one generation to the next — as in fact is the case in the Drosophila male — the phenomenon of inheritance would be very simple, but the orderly breaking apart of the linked genes in the female has made it possible to construct a theory as to the arrangement of the genes within each of the four chromosomes. Since the theory lends itself to numerical treatment it is possible to predict from a knowledge of the relations of a given gene to two other genes in the same group how it will be inherited with respect to all the other members of its group, which in three of the groups are over a hundred in number.

This practical aspect of the theory of linkage relations is from a theoretical point of view less important than the opportunity that the theory furnishes of giving evidence as to the arrangement of the fundamental elements of heredity, the genes, into systems of a higher order, the chromosomes. Were there no information as to the relation between the visible chromosomes and the linkage group it would still be possible to deal with the situation exactly in the same way by treating the linkage groups as a series of points held together in definite relations to each other. We might then speak of such groups as genetic chromosomes. But since there is abundant evidence at present to show that the linkage groups correspond to chromosomes it would be only pedantic to treat the linkage groups without respect to the chromosomes, i. e., to treat heredity as a purely formal mathematical problem. Moreover, quite aside from this aspect of the subject, the known behavior of the chromosomes has at times furnished the key to genetic problems, and, conversely, the restriction of this situation to the known limitations of chromosome behavior has kept the subject from expanding into unprofitable fields of speculation.

The orderly breaking up of the genes in blocks and the correlated interchange of equivalent blocks between the members of the two homologous series, or crossing-over, as it is called, has been described many times in the literature of Drosophila.

The demonstrations that the genes lie in linear order in the linkage group is most simply illustrated in a three-point experiment, i. e., one in which the three linked genes involved are so related that only single breaks occur between them. The complication introduced when two or more breaks occur in the series will be considered later. The case most often chosen for illustration involves three genes in the X-chromosome known respectively as yellow, white, and bifid.

Yellow and white are, according to the most recent calculations, 1.5 units apart. White and bifid are 5.5 units apart. If the order of the genes were yellow, white, bifid; then yellow and bifid should be 7 units apart (which is the actual result found), but if the order were bifid, yellow, white, then yellow and bifid should be 4 units apart. One or the other of these two relations is always found when the crossover values of three genes lying near together in the same chromosome (so that no double crossing-over occurs between them) are compared. This relation of the sum or the difference is spatially represented by three points in a line. It should be observed, here, that the accuracy of the result depends on the data being homogeneous; for, owing to variations in crossing-over (from external and internal factors) the results from different experiments may lead to discrepancies unless the proper corrections are made. CASTLE ('19*a*) fell into this error in his attempt to show that our data lead to the view that genes lie in three dimensions.

The conclusion arrived at from such a three-point experiment becomes more convincing when a fourth and then a fifth and sixth point is tested out against the first three points. In this way, provided sufficient points were available, the entire chromosome could be built up. As a matter of fact this is possible but scarcely necessary. It is not necessary because a knowledge of the proportion of double crossovers to single crossovers for each chromosome, or part of a chromosome, renders it possible to make due allowance for those regions when the genes are so far apart that double crossing-over is involved. In the X-chromosome it would be practically possible to construct its entire length on the basis of genes so near together that double crossing-over is excluded. Similarly for chromosome II. It would also be possible for chromosome III, except for its left end, where, fortunately, double crossing-over is low. In chromosome IV there is only about one unit between bent and eyeless; and shaven has given so little crossing-over with bent and eyeless that its position is undetermined, although, probably, it lies between the other two.

An illustration will serve to show how double crossing-over affects the ratios from which the calculations are made and to show also how it is possible to obtain the necessary corrections, that bring the results into line with those values obtained from single crossovers alone. For instance let us take three sex-linked genes: —

A fly of the triple recessive stock ruby (7.5), cut (20), vermilion (33) is crossed to a wild-type fly. The F_1 daughters (heterozygous for these factors, as shown in the diagram $\frac{\text{rb ct v}}{+\;+\;+}$) are crossed to triple recessive males of stock. The offspring fall into the following eight classes:

Non-crossovers		Single crossovers			Double crossovers		
rb ct v	539	in the first region	rb + +	137	in both regions	rb + v	11
+ + +	590		+ ct v	120		+ ct +	4
		in the second region	rb ct +	114			
			+ + v	107			

The total number of flies is 1622. Disregarding cut, the number of flies in which there is separation for ruby and vermilion is 478, or 29.4% of the total. However, ruby and vermilion are so far apart that not all the crossing-over between them shows itself in the classes that indicate upon inspection separation for ruby and vermilion, for when two crossings-over occur simultaneously between their loci the characters come out joined together as they went in. In calculating the total amount of crossing-over, account must be taken of this double crossing-over. The amount of double crossing-over can be measured through the presence of a character whose locus lies between the terminal ones, in this case "cut". The individuals that are ruby and vermilion but not cut, and the individuals that are cut but not ruby and not vermilion (see diagram) have resulted from double crossing-over, but do not themselves show separation for ruby and vermilion. They total together 15 (11 plus 4) which is 0.9% of all the flies. Since each of these flies represent two cases of crossing-over, the total number of crossings-over between ruby and vermilion is 29.4 plus 2 times 0.9, or 31.2 per hundred. This means that there were, on the average, 31.2 cases of crossing-over per hundred gametes.

In the experiment from which the preceding data were taken this is known to be all the crossing-over that occurred, because in the same experiment an additional character was present and followed, that lay between ruby and cut (namely crossveinless) and another (namely tan) that lay between cut and vermilion. No double crossing-over in either if these sections occurred (for full data see BRIDGES '20, page 662).

It is important to emphasize that the best data from which to obtain results that are more nearly comparable are those from a single female or from sister females reared at the same time under like conditions, i. e., in an environment as nearly uniform as possible. Even here care must be taken to avoid over-crowding, since this affects seriously the viability of certain classes. Such data are significant, especially for the relative cross-over values of the points involved, but are not necessarily the best from which to construct maps such as those we have published. These are constructed primarily to give the cross-over values that are most likely to be met with when no exceptional factors are present (Fig. 41). It is known, for instance, that certain stocks carry-factors that greatly affect crossing-over. Data from such sources are as far as possible to be excluded. It is known also that crossing-over changes with the age of the female. The data used are, as a rule, those obtained by the first ten day output. It is known that when the larvae are crowded certain classes are reduced. This complication is also to be taken into account.

It is important, therefore, that the data used in constructing maps should be carefully scrutinized and each lot given due weight. A detailed account of how the data have been handled is given by BRIDGES and MORGAN in the introduction of the paper entitled, "The Third Chromosome Group of Mutant Characters of Drosophila melanogaster," Carnegie Institution, Washington, pub. 327, 1923.

A caution must also be urged against a misunderstanding of the term "distance" as an interpretation of cross-over values. Strictly speaking the location of the loci on the map is an expression of the percentage of crossing-over between adjacent points. If these points be in linear order, as the analysis of the data requires if they are located in space, it follows that the percentages of crossing-over must be interpreted as distance; but the distances apart of the genes in the same series are comparable only if the crossing-over is as likely to occur at one level as at any other. If this were not so, the linear order might hold and the crossing-over values would be available for predicting untried combinations, but the spacing of the points on the map might be very inaccurate or at best only approximately true. The latter alternative expresses our attitude towards map distance. An example may

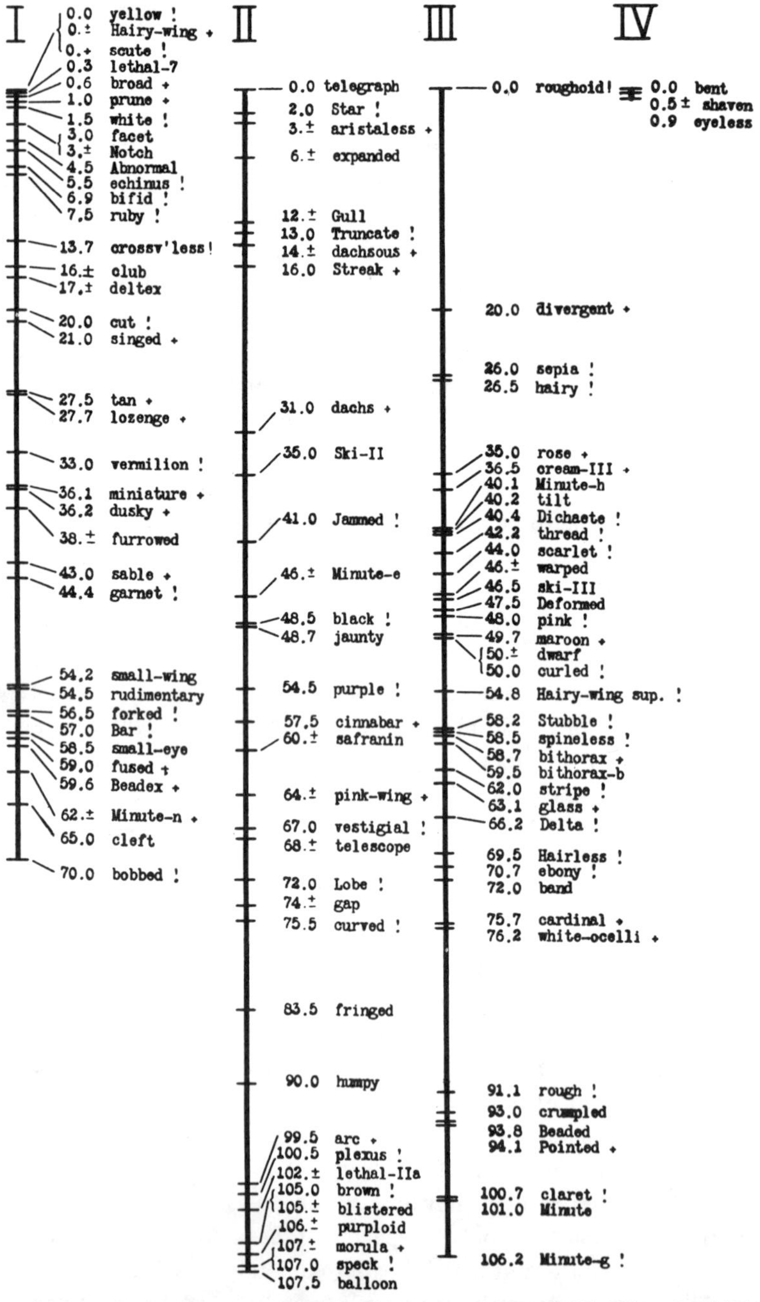

FIG. 41. Maps showing principal loci in chromosomes I, II, III and IV. The symbol ! designates the most useful types, +, those nearly as good, while those unmarked are important only in special connections.

serve to illustrate what has just been said. Suppose crossing-over between a and b is apparent in 10% of the offspring. Then a and b stand 10 units apart in the map. Now if some factor, either genetic or environmental should affect crossing-over so that it was reduced to 5% in the total of individuals, the distance between a and b would be expressed as only 5 units. Now if crossing-over is more frequent at the ends of the chromosomes than in the middle, as a constant feature of crossing-over, then the loci are relatively too far apart at the ends and too near together in the middle of the chromosome. That such regional differences exist is probable, as the analysis of certain data indicates. Hence the distances as they appear on the map are not to be taken too literally but only as approximations.

An understanding of these questions will remove the unwarrantable criticism of our interpretation of the localization of the genes that has been brought forward by DETLEFSEN ('20). He carried out with D. melanogaster an experiment in which, in each generation, individuals were chosen, as the parents of the next generation, from cultures in which a lower percentage of crossing-over took place. In a relatively short time the decrease in the crossover values appeared, but soon slowed down. From this evidence DETLEFSEN intimated that the location of the genes on the maps had no real significance. His criticism might have appeared to have weight had we not explained more than once what the map values represent when intrepreted as distance. DETLEFSEN'S criticism failed to bring out that we had already discovered and described both genetic and age and environmental factors that influence crossing-over. In fact STURTEVANT had carried out an involved investigation of a case in which a genetic factor profoundly affects crossing-over. He had shown that the factor involved was a genetic factor of some sort that could be dealt with as are other genetic factors. It is quite obvious from DETLEFSEN'S results also, that selection changed the crossing-over values by bringing together or else by removing genetic factors already present in the material. The results have, therefore, no greater value than in demonstrating another case of a change in genetic behavior by the isolation of genetic factors. The order of the genes involved in the selection experiment was in no way affected by the selection, and the theory of the order of the genes remained unaffected by this work.

It has been intimated that the absence of crossing-over in the male of Drosophila affects in some mysterious way the validity of the hypothesis of the linear order of the genes, but since it is a fact that such a difference between the male and female exists, it is not obvious to us that there is any basis for criticism on this score. It would of course be interesting to discover the differences in the sexes that lead to these results, but it is futile to speculate about them until cytology is in a position to help out the situation. In passing, it may be noted that there are on record one or two cases in Drosophila melanogaster where crossing-over may have occurred in the male; but these cases are so few and erratic that until all other possibilities have been eliminated they may safely be laid aside. There is, of course, no theoretical objection to the occurrence of crossing-over in the male; and even if it occurs, as it does apparently in other groups, there is nothing in the phenomenon to alter the theory of the linear arrangement. More interesting and significant are the numerous discoveries of cases in which crossing-over in one or another chromosome of the female is suppressed (see pages 100—108).

The lengths of the genetic chromosomes (see Fig. 41) are (I) 70, (II) 107, (III) 106, and (IV) 1 unit. Their relative lengths are roughly the same as the lengths of the actual chromosomes as observed in the metaphase plate. From this we conclude that the maps of the mutant genes cover approximately the entire length of the chromosomes. This implies that, on an average, a unit of crossing-over covers the same length of chromosome in all the groups.

Inspection of the maps (Fig. 41) shows that at the "left" (zero end) of chromosome I, and near the middle of chromosome III, and near the middle and at both ends of chromosome II, there are regions where mutant loci are especially numerous in proportion to the length of the section of map involved. This may mean that these crowded regions represent a relatively long section of the actual chromosome, one in which a unit of map-distance corresponds, in comparison with other regions, to an abnormally long section of chromosome.

Conversely, the empty regions, namely, the middle of chromosome I and the middle of the right and left limbs of chromosomes II and III, presumably represent relatively small sections of the chromosomes, or ones in which crossing-over is relatively frequent.

Linkage groups II and III correspond to the V-shaped chromosomes whose spindle attachment is at the apex of the V. The midpoint of the map of these chromosomes should be near the point of spindle-fibre attachment. Likewise the median locus in group II and III (since mutation is equally likely in the two halves of the V) should be near the spindle-fibre attachment. Judged by these two facts, and by the fact that many crossing-over modifiers apparently affect one half and not the other, this point of spindle-fibre attachment lies near purple in chromosome II and near pink in chromosome III.

The recent work on linkage relations in attached-Xs ("double"-Xs) indicates that the right end of the map of the X-chromosome is the end to which the spindle-fibre is attached.

Inspection of these maps shows no apparent relationship between the location of a gene and the part of the body affected by it, or similarity to neighboring genes. Thus, in the left end of the X-chromosome, in close proximity, are genes affecting eye-colors (white and prune), body-colors (yellow, straw-b, silvery, chlorotic), bristle-numbers (scute), head-size (giantoid), wing-shape (broad), number of hairs (Hairy-wing). In chromosomes I, II, and III there are genes that affect the same part of the body in closely similar ways, as, for example, the almost indistinguishable eye-colors vermilion (I), cinnabar (II) and scarlet (III). Again, sable (I) black (II) and sooty (III) are body-colors much alike. Garnet (I), purple (II), and maroon (III) are very closely similar. This evidence seems to mean that mutations whose loci are in the same region of the germ-plasm may not effect the same or neighboring parts of the body any oftener than would be expected if the genes were located at random. In a word, the pattern of the germ-plasm does not correspond to the pattern of the body.

Cytological basis for Crossing-over

That the separation of two sex-linked mutant characters, that entered a cross together, is brought about by some type of breaking apart of the X-chromosome and the replacement of the part carrying the gene for one of these characters by the corresponding part from the other X carrying the wild-type allelomorph of that character, became a necessary conclusion as soon as it was shown that the genes for the sex-linked characters were carried in the rod-shaped X-chromosome. The genetic evidence (from equational non-disjunction, from the 3N data, and from the attached-Xs data) has shown that this "crossing-over" occurs at a stage at which each parental chromosome has become split into two daughter chromosomes, i. e., at a four-strand or tetrad stage; and that the crossing-over involves two only of the four strands at any given point of crossing-over. A survey of the cytological observations on the formation of gametes in various plants and animals shows that the only stage at which such an interchange could occur regularly is between the last gonial division and the first maturation division, that is, in the "growth-stage". In *D. melanogaster* there is direct evidence that it is in this stage that the crossing-over does occur. The genetical evidence in crosses in which many characters are simultaneously showing the accustomed frequencies of separation rules out the hypothesis ("reduplication") that the separations are accomplished in gonial or pregonial stages (STURTEVANT '14). PLOUGH ('17) made an investigation of the length of time taken for an egg in passing through the growth stage, and showed that this corresponded to the length of time taken for a change in temperature to produce a change in the frequency of crossing-over, and concluded therefore that the crossing-over could not have occurred before the beginning of the growth stage. In Drosophila, cytological examination of the growth stages of the egg shows so little detail that nothing has been established thereby as to the nature of the process of crossing-over. Examination in other animals has yielded suggestive pictures. Thus, in Batracoseps, JANSSENS (LA CELLULE '09, '24) finds that, following the last spermatogonial division, the chromosomes spin out into fine strands that become disposed as loops with the ends close together at one pole of the nucleus. The ends of the homologues of each pair begin to unite into a thread of double thickness. At least in the large, easily observed chromosomes, this "synapsis" proceeds from both ends to completion in the middle of the loop. Then the double threads, of haploid number, begin to shorten and thicken, and presently to open out, whereupon four distinct strands can be seen. These four strands are often disposed in such a way that breaking and transjunction seems to have occurred between the two inner strands. In subsequent stages the chromosomes shorten still further into typical tetrads which enter the spindle of the first maturation division. Similar four-strand figures with possible transjunctions are seen in the grasshoppers (WENRICH '16, Bull. Mus. Comp. Zool. Harv. 60. '17, Jour. Morph. 29) and in other forms. These observations do not prove that crossing-over has occurred in the specific cases studied (ROBERTSON '16, Jour. Morph. 27), and in the case of the grasshoppers practically no crossing-over takes place in the male (NABOURS, unpublished data). However, the cytological observations have thus far not produced any evidence that cannot be harmonized with the rather rigid requirements that the genetic evidence imposes. (For an extended discussion see WILSON and MORGAN '20, Am. Nat. 55).

Coincidence

If three mutant characters a, b, and c have loci in that order in the X-chromosome, and a cross is made between a fly carrying all three genes and a fly carrying the corresponding wild-type allelomorphs, then any F_2 flies that show the characters a and c without showing b and likewise the flies that show b without showing either a or c must be accounted for by "double-crossing-over", that is, by a crossover between a and b, and a coincident crossover between b and c. When such crosses were made it was observed that the frequency of the doubles was generally much lower than was to be expected on the basis of the total amount of crossing-over between a and b and between b and c. Thus, if between a and b there were ten cases of crossing-over per hundred and between b and c 15 cases per hundred, the expectation for doubles is 10 percent of 15 percent or 1.5 percent. In such a case the observed percent of doubles would be about 0.3, or one-fifth of expectation. That is, the observed "coincidence" of double crossing-over was 0.3/1.5 or 0.2. In cases in which a and c were closer than about 15 units practically no doubles occurred. In cases in which a and c were very far apart as many doubles occurred as were expected. That is, the occurrence of a crossover at one point "interferes" entirely with the occurrence of another within a certain distance of the first, and interferes progressively less with second crossovers occurring at greater and greater distances from the first. The phenomenon of interference was first observed by STURTEVANT ('13*b*). Extensive studies by BRIDGES were generalized into graphs that showed for the X-chromosome the relation between distance involved and the corresponding coincidence (unpublished). MULLER ('15) made a study of the coincidences involved when the double crossing-over in the X was checked by mutants distributed throughout the length of the X. WEINSTEIN ('17) studied coincidence in cases in which one crossover occurred between two mutants located close together at the left end of the X and a second crossover occurred in specific sections to the right. WEINSTEIN'S data showed that beyond the initial stretch of zero coincidence, coincidence rose (interference decreased) until, at a point some 40 units from the first crossing-over, a second crossover was slightly more likely to occur than expected on chance; that beyond this apex in coincidence, coincidence again decreased (interference again increased) as far as could be followed. The decrease in coincidence beyond an apex indicates that the section included between the two crossovers of a double tends to be of a particular length.

BRIDGES (BRIDGES and MORGAN '23) showed that coincidence is a function of the particular chromosome and the particular section of each chromosome as well as of the length of chromosome involved. Thus, in the X, coincidence for a given length of map is greatest in the neighborhood of tan, in which region doubles may occur within a section as short as thirteen units. To the left of this region coincidences are lower and to the right they are lower still. In the second and in the third chromosomes the coincidences are highest near the mid-points (purple in II, and pink in III). Coincidence may be as high as 1.3 for the middle of III, and doubles occur within a section only six units long. From these mid-points coincidence falls off rapidly and symetrically to the right and to the left. Apparently the lowest point (interference highest) is some fifteen units from each end, where doubles are negligibly rare in a section as long as twenty units. The symmetrical distribution about the mid-point of chromosomes II and III is probably related to the fact that these chromosomes are V-shaped and the spindle-fiber attachment is at the mid-point.

JENNINGS (1923) has made a very careful and detailed mathematical analysis of the view that the linkage relations found in Drosophila are due to breaking and reuniting of linear series of genes. He concludes: "All the evidence is that the genes are in a linear series; the complex system of extraordinary relations shown by the crossover ratios is not explicable on any other ground" — a conclusion that most students of Drosophila regarded as already completely established. The assumptions made by JENNINGS give a very close fit to the data he used; but it should be pointed out that in one important respect his assumptions do not agree with the actual facts ascertained for Drosophila. He treats interference as being complete up to a certain point, and then from that point on, suddenly ceasing to act at all. The studies of MULLER (1916) WEINSTEIN ('18), and others show that in actual fact, interference is complete for a relatively short distance (10 to 20 units, depending on the region concerned), and then shows a gradual decrease, then still further away (in the case of the X of melanogaster at least) a second increase. While this discrepancy does not invalidate JENNINGS' conclusions from the data examined by him, it will apparently render his formulae useless for practical work with crossing-over and interference.

The relation between Crossing-over and the recombination of linked characters

In a previous section (page 88) the recombination of linked characters has been accounted for by the assumption of crossing-over of the chromosomes at some point between the two loci occupied by the genes of the characters. But when two crossovers occur between these loci the second restores the original combination of the genes (page 89), and thus for each such double, the number of recombinations of the linked characters per hundred individuals is less by two than the number of crossovers per hundred gametes. In the section just preceding this, it was shown that the relative number of double crossovers varies in a special way with the distance between the loci and also according to the particular part of the particular chromosome lying between the loci considered. Because of this last fact attempts to reduce the relation between frequency of crossing-over of the chromosomes and frequency of recombination of linked characters to a general formula must fail. This applies to the early graphs of BRIDGES (see page 97), to the generalization of HALDANE ('19), and to the calculations of JENNINGS ('22, '23). Special region-by-region studies of the different chromosomes, using many mutant characters simultaneously, are providing data from which the relation between recombination and crossing-over can be expressed as a series of graphs for any given chromosome. These graphs conform to all the special peculiarities of each chromosome and region as well as to the map-distance involved in each case. The graphs for the third chromosome have already been published (BRIDGES and MORGAN '23, also Fig. 42). A preliminary set of graphs for the X-chromosome, from unpublished data of BRIDGES, is given in Fig. 43. From the values given in these graphs, the coincidence expected for any given cross can be calculated.

Variations in Crossing-over

In working with the mutant characters of Drosophila it is observed that certain characters, e. g. miniature-wing and yellow body-color, are remarkably uniform, not being subject to noticeable fluctuating variation nor to frequent modification by specific modifiers. On the other hand, other mutant characters, e. g. truncate-wing and eosin eye-color, fluctuate markedly and are very frequently modified by other mutant genes. Similarly, in working with linkage, it has been found that certain regions of the chromosomes give remarkably uniform results, while other sections give great fluctuations within the same experiment, and are sensitive to environmental differences, and also give a great range of values according to genetic modifiers that are present frequently. Of the three large chromosomes the X-chromosome gives relatively very uniform results, the second less uniform, and the third quite divergent results. The mid-region of the third-chromosome is the most variable region, while the region from ebony to rough is fairly uniform in the frequency of crossing-over.

There is generally correlation between the variations in the different regions of a given chromosomes. Early in the work it was noticed (STURTEVANT '14) that unusually high or low crossing-over in one section of the X is usually accompanied by high or low crossing-over in other sections.

The most detailed study of the normal variations in crossing-over within an experiment in which conditions were supposedly constant was that of GOWEN on the third-chromosome (GOWEN '19). GOWEN found that within an experiment in which sepia Dichaete spineless sooty and rough were followed, the most variable region was between Dichaete and spineless; the two regions to the left and to the right (se-D and ss-e^s) were less variable, and the e^s-ro section was the least variable. The variations in the different regions were correlated. The highest agreement was between the mid-region and the region to the left of it. In comparing experiments involving a variety of third-chromosome genes GOWEN found that systematic differences of large magnitude were present, and that if data from all sources are treated together the remarkably high coefficients of variation show that crossing-over must be judged to be not a very precise process.

Besides the fluctuations and analysed variations from culture to cul-

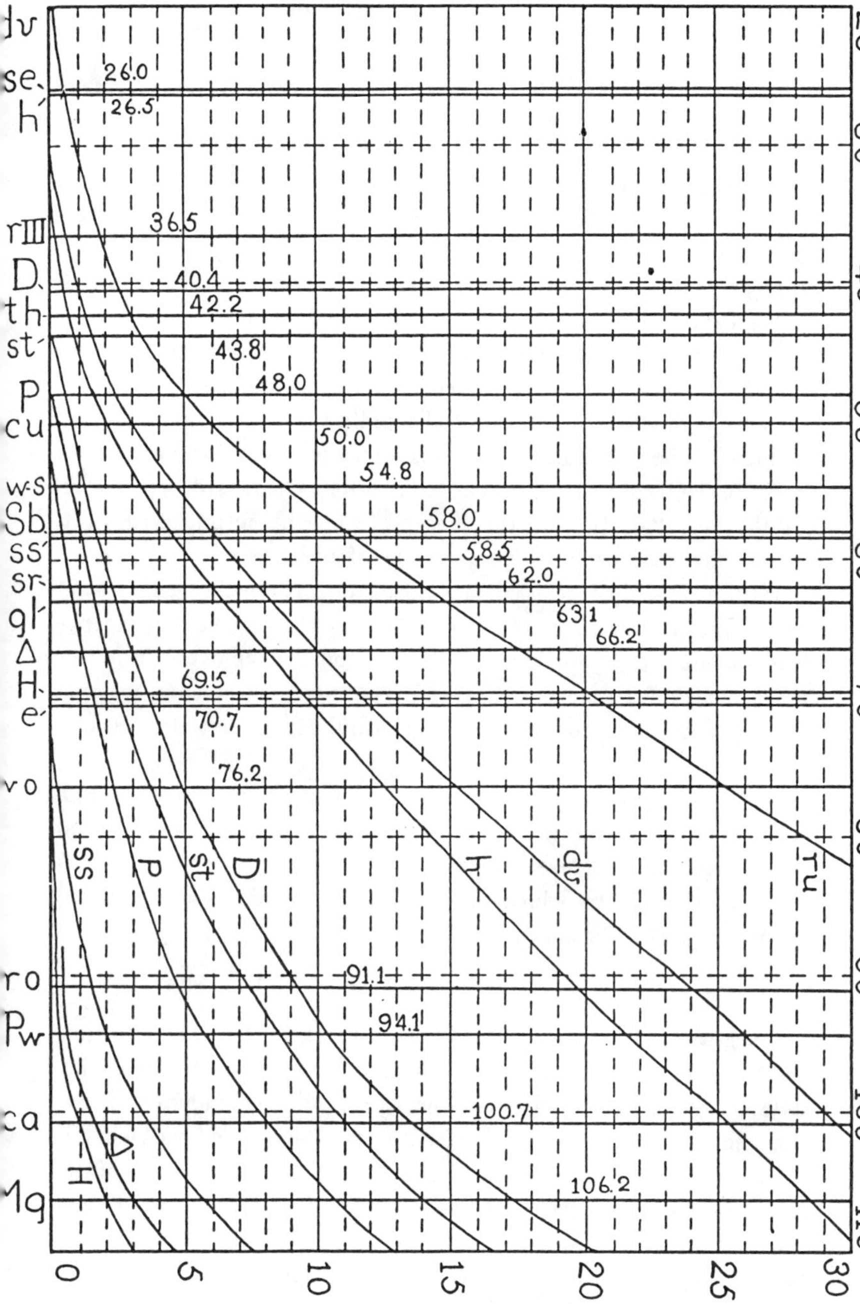

FIG. 42. Curves for chromosome-III, showing the amounts to be subtracted (ordinates) to convert cross-over values (abscissae) into recombination percents (see also insert in corner of Fig. 43).

ture within an experiment, it has been found that all the cultures, in experiments involving certain regions, show a systematic difference between the frequency of crossing-over shown by the eggs laid early and the eggs laid later. Thus, for the middle of the second-chromosome BRIDGES ('15) found that the eggs laid in a second brood, or ten-day laying-period, showed a combination percent only about 80 percent as great as that shown by first broods from the same mothers. Later experiments (by PLOUGH '17 and by BRIDGES), in which the eggs laid within successive two-day periods were compared, showed that for the mid-regions of the second-chromosome there was at first rather a high value, which dropped rapidly to a minimum about half as high by the eighth day, that a second rise reached a somewhat lower peak by the sixteenth day and a second fall reached a still lower level by the 24th. day. Beyond this length of time the output of eggs was too small and uncertain to give reliable information.

BRIDGES found that coincidence varied roughly inversely as crossing-over. Thus, for the series of two-day periods, the curve of coincidence was roughly the mirror figure of the curve of crossing-over. From this it was concluded that a change had occurred in the average interval between successive cross-overs as well as in the frequency of total crossing-over.

The early experiments on the change with age for the third-chromosome showed no significant difference between first and second broods. However, BRIDGES made an experiment using two-day periods and special stocks by which the third-chromosome could be checked for a great part of its length:

$$\frac{ru \quad\quad\quad\quad D \quad\quad p^p \quad\quad\quad\quad e^s}{\quad\quad h \quad\quad st \quad\quad ss \quad\quad} \text{♀} \times ru\ h\ st\ p^p\ ss\ e^s\ \text{♂}$$

The greatest change was for the scarlet-peach section, where the initial recombination percent was 7, which fell to a minimum of 2 at nine days, and then continued to rise throughout the remainder of the experiment, reaching 20 at twenty-eight days. The curves for peach-spineless, Dichaete-scarlet, spineless-sooty and hairy-Dichaete were of the same type as the scarlet-peach curve but of progressively less proportionate change. The roughoid-hairy curve was like the others for the first 18 days, but thereafter dropped to the end of the 28 day period covered by the experiment. Similar experiments for the first-chromosome did not show any significant effects of age upon crossing-over.

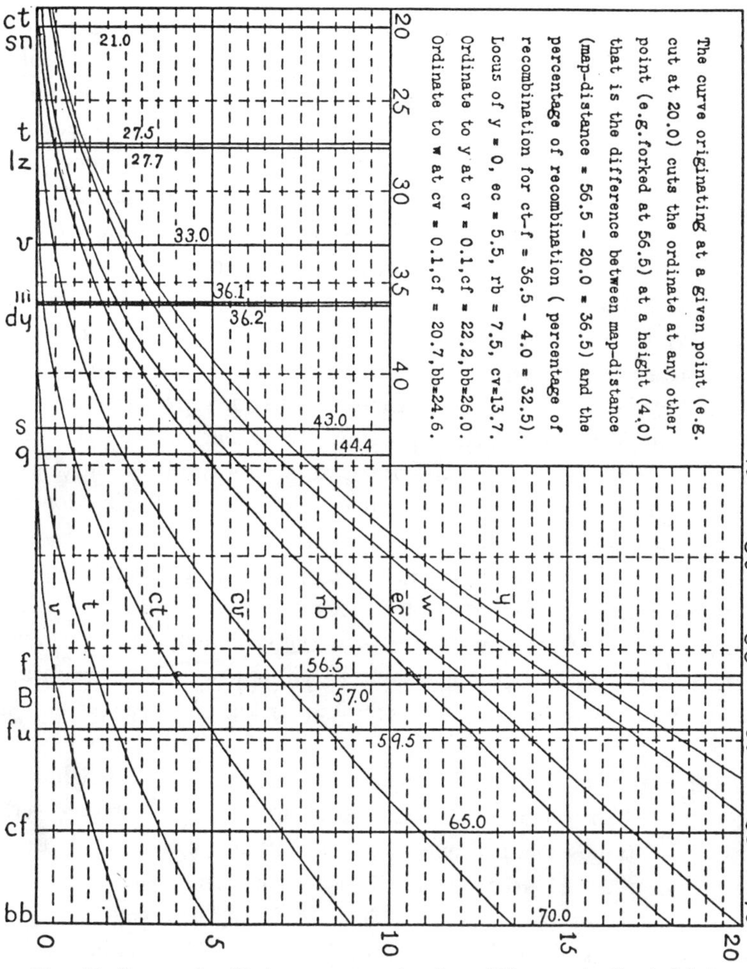

FIG. 43. Curves for X-chromosome, showing difference (ordinates) between map-distance (abscissae) and percentage of recombination.

These changes for the different chromosomes and regions are different enough from one another to show that the effect of age is not a simple one.

PLOUGH ('17*a*, '17*b*, '21) found, by comparing cultures raised at different temperatures, that there is a special relation between the amount of crossing-over and the temperature. The variation due to temperature was especially marked for the mid-regions of chromosomes II and

III and was negligible for the X-chromosome. Thus (see Fig. 41) for chromosome II, the amount of crossing-over between black and purple was practically the same for the whole range of temperatures at which flies are ordinarily raised, namely, 19° to 27°, but at 17.5°, on the one hand, and 29° on the other, there was a slightly greater amount of crossing-over. At 13° and at 31° the crossing-over reached an apex nearly three times as high as normal. Beyond these temperatures there were slight falls at 9° and at 32°, which are the extremes under which it is possible to raise flies successfully. By transferring flies to fresh culture tubes every two days, PLOUGH found that exposure to an abnormal temperature produced no effect upon the crossing-over observed until a period of about seven days had passed, whereupon the curve took an abrupt rise. This rise lasted approximately as long as the period of treatment; and the reversion to normal values was abrupt. From this it is plain that the developing germ-cell is sensitive to the heat action at only one period, rather early in its development. A comparison of the length of time between treatment and appearance of effect with the length of time taken for germ-cells to pass through the different stages indicates that this sensitive period is at the beginning of the growth period. From this it follows that any theory that places the mechanism for the recombination of linked factors earlier than this stage is untenable.

Neither age nor temperature change, nor any physical or chemical agent tried, had produced any significant change in the crossing-over for the X-chromosome. However, MAVOR ('23*c*) and MAVOR and SVENSON ('24) found that treatment by X-rays upon mothers heterozygous for eosin and miniature caused a decrease of crossing-over. With a treatment lasting about three minutes the reduction began about five days later, and dropped suddenly from the initial value of about 30 to about 10, at which value it continued with progressive but incomplete recovery from the sixth to the twelfth days. A dosage of the same total amount, but of lower intensity and longer duration, gave results approximately the same as the very short dose of high intensity.

A significant feature of these tests is that during the first five or six days the crossing-over is normal but the productivity is very low and the proportion of primary non-disjunctional exceptions very high. At this point the productivity and the proportion of exceptions become normal and the crossing-over becomes abnormal.

Similar experiments with black purple curved in the second-chromosome showed an increase in crossing-over instead of a decrease. This increase was greater for b-pr than for pr-c, and recovery was slower. This regional difference is roughly similar to that for age and temperature variations. The increase began at about the same interval after treatment (five days) as in the case of the decrease for the X-chromosome. It was highest for the next period, from the sixth to the ninth day, (b-pr rose from 5 to 15), and thereafter tended to become normal, but had not become completely normal by the eighteenth day.

The fact that a treatment produces an effect that begins after some five days suggests that the eggs already in the growth stage are not affected, as was also the case with the action of temperature. But that a treatment lasting only three minutes should produce an effect that lasts for at least twelve days, only gradually decreasing in amount, suggests that all the eggs in the oögonial stage are affected by the action, but that by the time the younger oögonia have reached the stage at which crossing-over occurs the processes going on in the cell have nearly recovered their normal course.

Besides the unanalysed variations in crossing-over within many experiments, and the systematic differences known to accompany changes in age and in temperature, and to result from X-rays, and possibly from other environmental differences, there are differences in the amounts of crossing-over corresponding to chromosome complexes different from the standard. The most familiar and the earliest discovered of these changes is the difference between the crossing-over in the 2N female and in the male, the male differing by the substitution of a Y-chromosome for one X. In the male there is no crossing-over between the X and the Y, nor between the members of the pairs of autosomes (MORGAN '12*g*, '14*e*). It is probable that this lack of crossing-over in the male is due to a general difference in the course of the early growth stages, particularly synapsis. But in the case of the Y there is an additional internal difference; for, when the Y is present in an XXY female, in which crossing-over is known to take place, the Y does not show crossing-over with the X (BRIDGES '16).

The presence of a Y in a female (XXY) results in two variations in crossing-over, namely, the XX eggs are always non-crossovers, and the amount of crossing-over shown by the X and XY eggs is slightly increased (BRIDGES '16).

In the case of the sectional deficiencies crossing-over is eliminated entirely throughout the length of the deficient sections. This is true of f B-deficiency (BRIDGES '17), vermilion-deficiency (BRIDGES '19), the Notch-deficiencies (MOHR '19, '23*d*), and the second-chromosome deficiency resulting from Pale-translocation (BRIDGES and MORGAN '23). There are several lethals, that are probably deficiencies, that involve decrease in crossing-over in the neighborhood of the lethal or even in more distant regions. Among these are lethal-2, lethal-IIa, and dachs-lethal.

Two cases of duplication are accompanied by crossing-over reductions in the neighborhood of the attached sections. Thus, the translocated piece of the second-chromosome greatly reduces crossing-over to the right of stripe. L. V. MORGAN has a case (unpublished) in which a piece of the left end of the X-chromosome (covering yellow, scute, broad) is attached to the right end of an X-chromosome. In a female having one such duplication, crossing-over between forked and the end is almost eliminated, and the alteration gradually diminishes to the left.

Finally, in 3N, the studies of BRIDGES and ANDERSON show that crossing-over in the X-chromosome is approximately doubled in the region to the left of ruby, and halved in the region to the right of ruby.

Variations in crossing-over due to genetical causes are very common, and account for many of the differences commonly observed within an experiment and in comparing different experiments. The first crossover variation of large magnitude having an analysed genetical cause was that of CIIL-CIIR, found by STURTEVANT in working with the second-chromosome (STURTEVANT '14, '19). As the name implies, this is a double modifier. A dominant crossing-over modifier (hence the symbol C) is situated in the second chromosome (hence CII), in the left limb of the second-chromosome (hence CIIL), where it eliminates practically all crossing-over from Star to purple, inclusive. Likewise CIIR, which was found in company with CIIL, is a dominant in the right half of the second-chromosome, where it eliminates crossing-over from purple to the end. When both are present there is a very little crossing-over in each limb of the chromosome, and a very little more in the neighborhood of purple.

The next crossing-over modifier was CIII, first encountered by STURTEVANT in some pink-ebony back-crosses (STURTEVANT '13), but not recognized as such until MULLER found the same modifier in a "spread" stock, and also in the balanced stock of Beaded. Later BRIDGES found it in a stock of maroon dwarf and also in a stock of eosin. It is not certain that those various strains of CIII are descended from a common stock in which the mutation to CIII occurred, and it is rather improbable that such is the case. CIII is a dominant located in the right limb of chromosome-III (near ebony) and it eliminates very nearly all crossing-over in the right limb from the end, up to a point just to the right of pink.

For the X-chromosome several modifiers are known. In working with high non-disjunction BRIDGES found that the stocks used were homozygous for a recessive crossing-over variation (cIa) that reduces crossing-over between vermilion and eosin to about 10 and between vermilion and forked to about 15. Thus, the action is stronger upon the left end (where the main factor seems to be located near ruby) and is less effective in the right end. There seem to be several autosomal factors that help cIa in producing this effect.

DETLEFSEN (DETLEFSEN '20, DETLEFSEN and CLEMENTE '23, DETLEFSEN and ROBERTS '21) in the course of some selection experiments on crossing-over in the X-chromosome, found that selection for an increased amount of crossing-over between eosin and miniature was ineffective. Races giving markedly lower percentage were obtained, one of these giving almost no crossing-over. No complete genetical analysis has been reported, but it appears that at least one major factor, and probably several minor factors were concerned.

MULLER (unpublished results) found that in an X-chromosome, containing the mutant genes scute, tan ^{2}, vermilion, small-wing and Bar, a mutant change occurred that eliminated crossing-over almost completely from the X-chromosome. The few crossovers that occur are mainly doubles in the mid-regions of the X. The modifier is in the X-chromosome, probably in the right end-region.

L. V. MORGAN (unpublished results) has found a similar crossing-over suppressor for the length of the X.

Besides CIIL and CIIR, STURTEVANT ('19*a*) found a modifier (CIII,II) located in the right limb of the third-chromosome, and affecting it, but also affecting the crossing-over in the middle of the second-chromosome. BRIDGES found a similar modifier (CII,III) located in the left limb of the second-chromosome, where it eliminated all crossing over between Star and black, but also reduced the crossing-over in the left limb of the third-chromosome.

Mrs. WARD ('23) found a dominant wing-mutation "Curly", with which is associated a CIIL and a CIIR. A curious feature of the CIIR in Curly is that a temperature of 30° C. increases crossing-over from the normally negligible amount to about 5% for the right limb of the second-chromosome.

Some stocks of cinnabar, sent by CLAUSEN (CLAUSEN '24) proved to have a dominant modifier that reduced the black-purple crossing-over from about 6% to about 3%.

The third-chromosome has the largest number of modifiers. We have already mentioned CIII, CIII,II, and CII,III.

PAYNE ('20) reports a dominant modifier affecting both limbs of the third-chromosome. Subsequent analysis (paper in press) showed that it was a double modifier (CPL, and CPR). When both are present there is a little crossing-over in the neighborhood of pink and very rare cross-overs elsewhere.

BRIDGES found a recessive eye-color, orange, located in the right limb of the third-chromosome and associated with a CIIIL and a CIIIR, which again allow a little crossing-over near pink.

L. V. MORGAN found a further similar case of a CIIIL associated with a CIIIR.

BRIDGES found three other CIIIR's; namely, "CIIIM", "CIIIN" and CIIIh.

The amount of change due to the modifiers so far mentioned has been so great that there has been little chance of failure to exclude the experiments in which they were present from the data to be used in constructing a "standard" map of the third-chromosome. Two modifiers found by BRIDGES are especially troublesome because of their overlapping the normal values, and because they are apparently widely distributed. One of these modifiers increases the hairy-scarlet value from about 14 to about 22. The other decreases the Dichaete-scarlet value from about 3.5 to about 1.5, and affects neighboring regions similarly but not to such a great proportional extent.

GOWEN (GOWEN & GOWEN '22) reports a recessive modifier located in chromosome-III but suppressing all crossing-over in all of the chromosomes.

CHAPTER XI

NON-DISJUNCTION OF THE X-CHROMOSOMES

The account up to this point has been an outline of the normal situation and relations present in *Drosophila melanogaster*. In the experiments upon which the foregoing account is based approximately 10,000,000 flies have been raised in pedigreed cultures, carefully examined, classified, and recorded. These normal relations are therefore thoroughly established and very familiar, and any deviations from the standard types of result are strikingly apparent. But more important still, the methods, tests, and breeding procedures have been constantly developing through experience with about 400 generations. We are now in position to lay hold of any new abnormal situation that arises, to breed the flies in such a way as to keep the abnormality from being lost, and then at our leisure to analyze its basis. In the present section an account will be given of different types of chromosome configurations departing from standard, with the methods by which each new situation was analyzed. It should be borne in mind that while these new types are usually referred to as "aberrations", in each case a new and self-consistent system of heredity comes to be established, with specific laws conforming to the new physical basis — the derived chromosome group.

The general classification and plan of treatment of chromosome aberrations that follows is based on the *unit of change*, which may be a whole chromosome, or a section of chromosome, or a group (n) of chromosomes [1]).

It is simpler to deal first with the changes in which a single whole chromosome is the unit of change, and then to examine similar pro-

---

[1]) See paper by Bridges ('23b) for extended classification of chromosome aberrations.

cesses in which the unit of change is a section of a chromosome or a whole group of chromosomes.

MUTATION

UNIT	PROCESS	RESULT
Gene	Transgenation ($G_0 \rightarrow G_1 \ldots G_1$)	
Section	Deficiency	n — a section } gamete
	Duplication	n + a section } gamete
	Translocation	New linkage relation
	Inversion	New linkage relation
Chromosome	Non-disjunction	n + 1, n — 1 } gamete
	Elimination	n — 1 } Mosaics
	Compounding	n→n′, where n′ < n
	Fragmentation	n→n′, where n′ > n
Group (n) nucleus	Nuclear non-division	2n } gametes
	Nuclear fusion	2n } gametes
	Binucleation	Mosaics
	Dispermy	Mosaics

The work on non-disjunction of the sex-chromosomes has been reported in a series of papers (Bridges '13, '14, '16, Safir '20, Mavor '21*a*, '21*b*, '22, '23, Bonnier '22). Some unpublished work by Bridges will be included in the following account.

The phenomenon of non-disjunction in *Drosophila melanogaster* was first detected (1911) through the rare occurrence of an individual that broke the rule for the transmission of sex-linked characters. The rule is (see page 87) that when a female homozygous for a sex-linked recessive is outcrossed to a male carrying the wild-type allelomorph, all the sons are like the mother, and all the daughters are like the father (criss-" cross" inheritance). The explanation advanced by Morgan to cover this type of inheritance, postulated that the sons were like the mother with respect to all sex-linked characters, because the genes of the sex-linked characters were carried in the X-chromosome, and that the single X-chromosome carried by each son was derived from the mother. This hypothesis could not account for the observed occasional occurrence of an exceptional red-eyed son among the regular white-eyed sons from the cross of a white-eyed female and a red-eyed male [1]. Nor could it account for the (still rarer) occurrence of an exceptional white-eyed daughter among the regular red-eyed daughters of the same cross.

The hypothesis advanced by BRIDGES to account for these exceptions was that:

(1) the genes of the sex-linked characters are carried by the X-chromosomes, but that

(2) the X-chromosome of the exceptional red-eyed son was derived from the red-eyed father, and that

(3) both the X-chromosomes of the exceptional white-eyed daughter were derived from the white-eyed mother. He further assumed that

(4) the exceptional types of sons and of daughters were both products of a single abnormal process, namely, that

(5) *occasionally the two X-chromosomes of the white-eyed mother failed to disjoin from each other at the reduction division, but, instead, both remained in the egg or both passed out to the polar body.*

In case both passed out to the polar body, an egg without a sex-chromosome, but with one of each of the autosomes, would be produced. This egg, if fertilized by an X-sperm of the father, should give rise to an individual that should be a male (having only one X-chromosome) and that should be red-eyed like the father (this X-chromosome having been derived from the red-eyed father).

Likewise the XX-egg (two-X egg), fertilized by the no-X sperm [2], should give rise to an individual that should be a female (having two X-chromosomes) and that should be white-eyed like the mother (both X-chromosomes having been derived from the white-eyed mother).

When the exceptional white-eyed daughters were outcrossed to wild males with red-eyes, it was found that further exceptional daughters and sons were produced. To account for this, it was assumed (BRIDGES '13, p. 504) that the X-chromosomes of the original white-eyed mother had contained a gene that predisposed the X-chromosomes to non-disjunction, and since the exceptional daughters contained these same

---

[1]) Such exceptions were also found to occur in crosses in which a great variety of sex-linked characters was involved (BRIDGES '13), but for simplicity the present discussion will be restricted to the most familiar of the sex-linked characters — white eye-color.

[2]) At that time the Y-chromosome had not been clearly identified, and the constitution of the male was supposed to be X O.

X-chromosomes, they should contain these same genes and likewise be predisposed to give non-disjunction.

The above hypothesis was unsatisfactory in one respect — that the percentage of exceptions in the derived cultures was markedly higher than in the original cultures. In the secondary cultures the number of exceptions was roughly one in 20, (see page 118) while the original exceptions had been an isolated individual here and there, an incidence in the neighborhood of one in 2000 individuals (see page 113).

Furthermore, in spite of the very extensive work of STEVENS ('08), it was still uncertain whether the male *Drosophila* was to be regarded as of the XO or of the XY-type. If a Y-chromosome were normally present, it should be transmitted through half the sperm (that half formerly called simply no-X sperm), and should be present in each of the exceptional daughters (XX egg by Y-sperm, instead of XX egg by no-X sperm).

It was then suggested that a Y, present in these exceptional females (XXY), might itself be the cause of their producing secondary exceptions, and that the assumption of a gene favoring non-disjunction was erroneous. The correctness of this view with respect to the Y has been established by a variety of genetic tests and by cytological examination of the different classes, as will be shown in the section dealing with secondary non-disjunction.

We may now define as primary exceptions those occurring when the mother is XX, and not XXY, in constitution. In many cases where exceptions are first encountered in an experiment it is difficult to be sure from the pedigree whether the mother was XX or XXY, but the difference in the percentage of the exceptions (roughly 1 in 20 or 1 in 2000) is usually decisive. A close calculation of the percentage of primary exceptions can of course be made only from those cases in which the constitution is known from the pedigree, or from special tests carried out on the offspring.

The cultures of BRIDGES that met the above requirements gave a total of 20,484 flies, of which 12 (or 1 in 1700), were exceptions.

Later, SAFIR ('20) carried out a large series of tests of the frequency of primary exceptions in various stocks (Lines 2—6, table 3).

The control experiments made by MAVOR in connection with his X-ray work gave some 38,000 further individuals (table 4). The total data included more than 200,000 flies, of which roughly 1 in 1200 were exceptions. However, one of the series of experiments, that in which eosin miniature XX mothers were used, gave markedly more exceptions (1 in 500) than the others. That this difference was due to a recessive gene that favored the production of primary exceptions is strongly suggested by the fact that the F_1 daughters from the outcross of eosin miniature females to wild males, (the w^e m, F_1, of table 3) did not give the high ratio of exceptions. Accordingly, the eosin miniature series of experiments may be excluded on the ground that it belongs to a different genetic category from the rest; and the ratio calculated from the remaining 170,000 flies, namely 1 in 2000, is probably a better standard of expectation as to the normal frequency of primary non-disjunction.

The observed frequency of primary exceptions is half that of the occurrences of non-disjunction, for it is just as probable that a given no-X egg is fertilized by a Y-sperm as by an X-sperm. A no-X egg by a Y-sperm gives a non-viable YO zygote, and thus half the cases are lost to observation. Likewise the half of the XX eggs that are fertilized by X-sperms result in XXX zygotes that ordinarily perish.

TABLE. 3. SUMMARY OF DATA UPON PRIMARY NON-DISJUNCTION.

	Reg. ♀	Ex. ♀	Ratio	Reg. ♂	Ex. ♂	Ratio	Tot. Reg.	Tot. Ex.	Ratio
BRIDGES '16	10,635	5	2100	9,849	7	1400	20,484	12	1700
w^em, SAFIR '20	25,021	17	1500	22,524	70	300	47,545	87	500
v, SAFIR '20	12,350	2	6200	10,638	17	600	22,988	19	1200
w, SAFIR '20	11,310	2	5600	10,463	13	800	21,773	15	1500
w^em F_1, SAFIR '20	24,433	4	6100	23,757	14	1700	48,190	18	2700
v F_1, SAFIR '20	9,625	2	4800	9,417	10	900	19,042	12	1600
MAVOR, table 4	17,103	2	8006	20,842	11	1900	37,945	13	2900
Tot. (excluding w^em)	85,456	17	5000	84,966	72	1200	170,422	8	2000

Early in SAFIR's experiments it was noticed that the primary male exceptions were roughly four times as numerous as the primary female exceptions. The total data (including w^em) gave 142 male exceptions to 34 female exceptions. This preponderance of male exceptions can be formally accounted for by supposing that in primary non-disjunction the two X's both pass into the first polar nucleus more often than they remain in the egg nucleus. Such a preferential direction to the first polar spindle seems an arbitrary assumption, especially since in secondary non-disjunction there is no preferential direction for the XX-Y and the XY-X reduction spindles (pages 118 and 119). It seems better to assume that the numbers of XX and of zero eggs produced by simple primary non-disjunction are equal; the excess of zero eggs may then be supposed to be produced by a modification of the non-disjunctional process, in the following way.

It seems probable that primary non-disjunction is initiated by some unusual local difference, in consequence of which the synapsed X's lag upon the reduction spindle and separate from each other only with difficulty. The chemical and physical forces involved in the attraction and repulsion of homologous chromosomes may be affected by some environmental difference; or the delay in the separation of the synapsed X's may be due to a purely mechanical entanglement; the two X's may have become intertwined in such a fashion or to such a degree that disjunction is more than ordinarily prolonged. If this delay is not too great the disjunction may be accomplished, and each daughter nucleus receive its X. With a slightly greater amount of delay in separation both X's may be caught in the nearer nucleus. In case this nearer nucleus is the first polar nucleus, a no-X egg results that gives rise to a male exception (no-X egg by Y-sperm). In case the nearer nucleus is the egg nucleus a two-X egg results that corresponds to the exceptional daughters. There seems no reason to suppose that these two classes should be unequal; and thus simple primary non-disjunction should give an equality of male and female exceptions. The frequency of simple primary non-disjunction is then shown by the frequency of female exceptions, which (see table 3) is observed to be roughly 1 in 5000. The modification referred to above may be supposed to happen when the synapsed X's lag to a still greater extent, and are left in the middle of the spindle when the daughter nuclei form; i.e. they do not get into either nucleus. Only no-X eggs result from such reductions, and consequently only male exceptions come from such a process, and give an excess of male exceptions.

Male exceptions produced by primary non-disjunction should be XO in constitution (no-X egg by X-sperm) that is, they should be without a Y-chromosome. The cytological examination carried out by SAFIR, (SAFIR '20) showed that they are in fact without a Y-chromosome.

A comparison of the characteristics of these XO males with normal XY males shows that the external and somatic characteristics of an

XO male are not distinguishably different in any respect from those of a normal XY male.

Moreover, when XO males show various mutant characters, these have not been found to be modified by the absence of the Y-chromosome. The conclusion is therefore that the Y-chromosome is deficient in genes, and specifically is deficient in the genes that are carried by the X, its mate.

However, there is one striking pecularity of XO males. They are invariably and entirely sterile. BRIDGES tested about 15 such males, and SAFIR more than fifty others. XO males do not differ in any visible sex-characteristic from XY males. They have normal external genitalia, and they court and copulate in the normal manner. However, the spermathecae and oviducts of females that have copulated with XO males contain no sperm. The fluids in the male ducts contain spermatozoa, but in small quantity. They are non-motile. In the tube of the testis few bundles of sperm are present and these do not readily break up into individual spermatozoa. It was found that the testis contained spermatogonial cysts that were syncytial in nature. They consist of large rounded masses of cytoplasm containing many nuclei. Probably this accounts for the later imperfect development of the spermatozoa. The maturation divisions were not entirely normal, but evidently the abnormality that results in sterility begins before the maturation stages are reached, and the defectiveness of the sperm is not due to the disturbance of the mechanism of maturation through the loss of a gear wheel, but rather to a general change attributable to the action of the Y upon the development of the testis and spermatogonial tissue, that is, to genes of the Y-chromosome.

If primary non-disjunction occurred in the XY male, XY and zero sperm would be formed. Since the zygotes from these sperm would not differ in their sex-linked characters from regular offspring such an occurrence could not be detected in the immediate offspring. However, the XY sperm would give rise to XXY daughters which would then give rise to secondary exception, and the zero sperm would give rise to sterile males. Neither of these types of offspring could ordinarily be distinguished from like types originating from non-disjunction in the female. A decisive proof of the occurrence of primary non-disjunction in the male has been found in the course of studying secondary non-disjunction in the female. Four XX daughters have been found, each of which received both her X-chromosomes from the XXY mother, and neither an X nor a Y chromosome from the XY father (unpublished work of BRIDGES).

In a series of papers ('21*a*, '21*b*, '22, '23) MAVOR reported that treatment of females by X-rays caused a marked increase in the production of primary exceptions (table 4, including some unpublished data communicated by letter). The treated females produced 222 exceptions and 27,233 regular offspring, or approximately 1 in 120. The untreated sister females produced 13 exceptions and 37,445 regulars, or 1 in 2900. That is, the exceptions produced by treated females were 25 times as numerous as those produced by untreated sisters.

The ratio of female to male exceptions (33 : 189 or 1 : 5.7) shows an even greater excess of male exceptions than does standard primary nondisjunction (1 : 4, see p. 113).

Four of the sixteen female exceptions of MAVOR'S 1923 data were "equationals" (see p. 126), the other twelve were of the usual "reductional" type.

Nearly all of the exceptions produced by the X-rayed females were from eggs laid during the first six days after raying. During this first period (1) the number of offspring was a small fraction of the normal productivity, (2) the proportion of non-disjunctional offspring was very high, and (3) crossing-over relations were normal. At the end of the six-day period there was a rather abrupt change in all these respects: (1) productivity rose to normal, (2) the production of exceptions fell to normal, and (3) the crossing-over frequencies became abnormal (see p. 100 for full discussion of change in crossing-over).

The magnitude of all these effects was more or less proportional to the degree of the treatment, but a given quantity of rays produced the same effects whether administered at low concentration over a period of many hours or at high concentration for a few minutes.

Since, according to PLOUGH ('17), the development of the egg from oogonial stage until laying, occupies about six days, it was inferred that the X-rays affected all the eggs that were then in the growth stage, and in such a fashion that (1) they were rarely viable, and that (2) the maturation mechanism was disturbed. The fact that both of these effects disappeared sharply, led to the inference that the cells that were in the oogonial stage were not affected in those respects. The effect upon crossing-over must be attributed to a change upon the oogonial stage and not upon the growth stage. The effect upon the oogonial stage is a reversible one, since the change in crossing-over gradually approaches normal after the initial departure.

TABLE 4. — SUMMARY OF DATA FROM MAVOR ON PRODUCTION OF PRIMARY NON-DISJUNCTION BY X-RAYS

Source	Mating	X-rayed mothers				Control mothers			
		Reg. ♀	Ex. ♀	Reg. ♂	Ex. ♂	Reg. ♀	Ex. ♀	Reg. ♂	Ex. ♂
Mavor '21b	+ ♀ × w ♂	—	—	1,421	24	—	—	3,734	1
Mavor 22	w ♀ × w^em ♂	510	2	455	12	1,743	—	1,725	1
Mavor '23	w/w^em ♀ × + ♂	3,944	16	3,491	84	12,696	2	12,640	8
Unpublished	w/w^em ♀ × + ♂	8,609	15	8,803	69	2,664	—	2,743	1
Total.		13,063	33	14,170	189	17,103	2	20,842	11

Secondary non-disjunction

On page 112 it was stated that the cause of the production of exceptions by mothers, that were themselves exceptions, is the presence in those mothers of an extra Y-chromosome. The proof for this came from both genetical and cytological investigation.

The genetic tests were as follows:

(1) A linkage experiment was made (BRIDGES '16, page 45) in which an assumed sex-linked gene for non-disjunction was found to show no linkage to two known sex-linked genes, eosin and vermilion.

(2) A stock (obtained according to a rigorous method that should yield a stock homozygous for any given sex-linked gene, and hence for a gene for non-disjunction if such existed) was found, upon testing, not to be homozygous for non-disjunction. (BRIDGES '16, page 49).

The cytological tests were as follows:

(1) An examination of the chromosomes in the spermatogonia of the male was made which showed clearly that there is a Y-chromosome in the male. This Y-chromosome is a J-shaped chromosome, slightly longer than its mate the X, which is rod-like. In the female, as STEVENS showed, there are two rod-like chromosomes, but there is no Y-chromosome. (BRIDGES '16, page 110, and unpublished material. See also page 112).

(2) In the breeding experiments it was found that half of the regular daughters of an exceptional daughter, gave each about 5 % of secondary exceptions, and that the other half of the regular daughters gave no exceptions. It was, therefore, predicted that, if secondary non-disjunction were due to the presence of an extra Y-chromosome, cytological examination of the regular daughters should show that half of them carried an extra Y and half did not. Good metaphase division figures were found in 18 regular daughters. Nine of these carried an extra Y and nine did not.

In both primary and secondary non-disjunction it has been shown that only sex-linked characters have been transmitted in an abnormal fashion; that is, only the sex-chromosomes and not the autosomes are involved.

Numerous other genetical tests and cytological checks have demonstrated the correctness of the view that the Y is the cause of the production of secondary exceptions. These tests will be treated in subsequent paragraphs.

On page 112 it was stated that in secondary non-disjunction the frequency of exceptions is much higher than in primary non-disjunction. A summary of the cultures known to be secondary in the 1916 paper of BRIDGES gives a total of 58,037 flies of which 2481 or 4.3 per cent were exceptions. The percentage of exceptions among the females was 4.3, precisely the same as among the males (table 5). This equality proves that an XXY female produces XX and Y eggs in equal numbers, that is, that in XX—Y reduction it is a matter of chance whether the egg receives the XX or the Y end of the spindle (see Fig. 44).

TABLE 5. — SUMMARY OF DATA ON FREQUENCY OF SECONDARY EXCEPTIONS

Source	Reg. ♀	Ex. ♀	% Ex. ♀	Reg. ♂	Ex. ♂	% Ex. ♂
BRIDGES '16, p. 15	27,679	1225	4.3	26,391	1169	4.3
" '16, " 16				1,496	77	4.9
Totals	27,679	1225	4.3	27,887	1246	4.3

When the regular daughters of XXY females were outcrossed, it was found that half of them gave exceptions and half did not (80 XXY: 79 XX, BRIDGES '16 p. 27). That is, the extra Y remains in the egg or passes out into the polar body at random. There should likewise

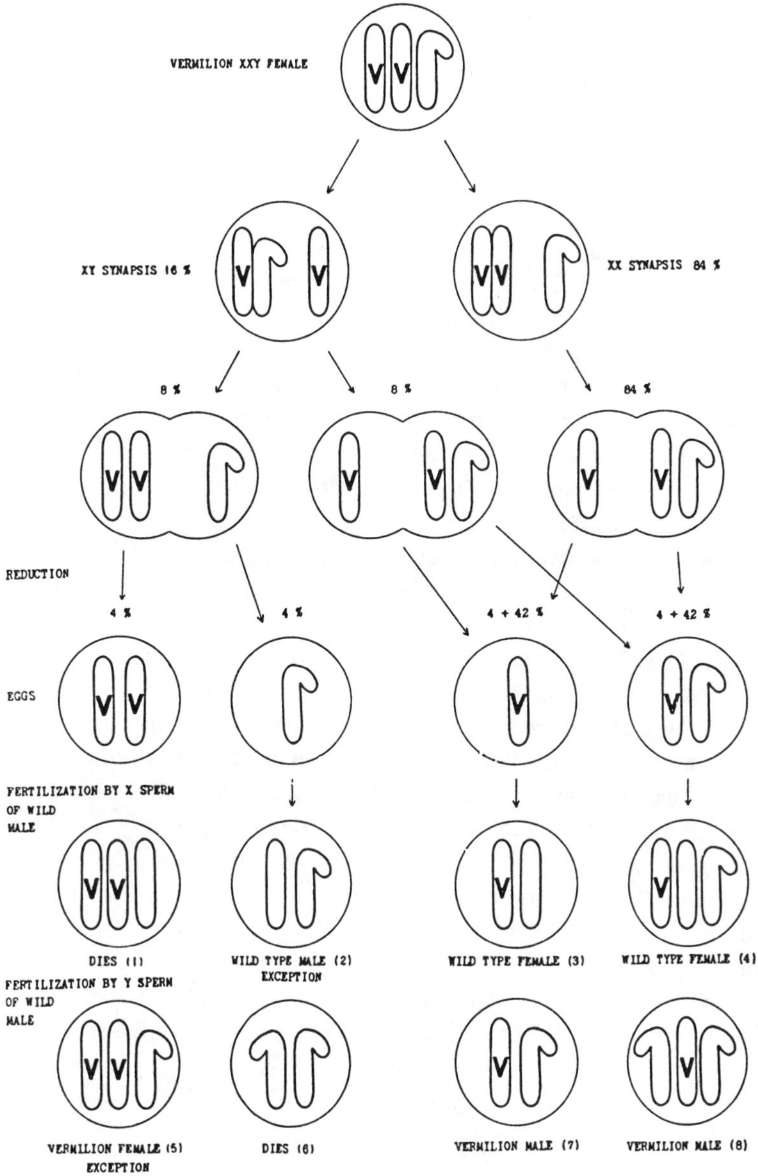

FIG. 44. Secondary non-disjunction in a vermilion XXY female, mated to a wild male.

be two types of regular sons, viz. XYY, and XY. Of the fourteen regular sons tested, nine were XYY and five were XY (BRIDGES'16, p. 37).

Linkage experiments have shown when an ordinary XX female carries the recessives eosin and vermilion in one X, and forked in the other, that about 48 % of the X-chromosomes that emerge have undergone crossing-over. But when an XXY female carries eosin and vermilion in one X and forked in the other, it was found that the exceptional daughters invariably likewise carried eosin and vermilion in one X and forked in the other; that is, none of the X-chromosomes present in the XX eggs had undergone crossing-over. The same fact, that exceptional daughters of XXY mothers are always non-crossovers for both the X-chromosomes, was observed in a great variety of other crosses (BRIDGES '16, p. 114).

In contrast to the total lack of crossing-over shown in the X's that descend to exceptional offspring, the X's that descend to regular offspring (X and XY eggs) have undergone the standard amount of crossing-over or probably even more. Table 6 is a compilation of the data from BRIDGES ('16) in which the amount of recombination in cultures where the mother was XXY can be compared with the amount shown in cultures from full sisters that were XX. In all cases there was somewhat more recombination in the XXY cultures than in the XX controls. If each percentage increase is weighted according to the sum of the flies on which the calculation was made, the weighted average increase over the controls is 5.2 %. But an examination of the different cases suggests that perhaps they are not all members of the same system. Thus, it is perhaps true that there is a gradient of increase from the left end to the right. The left-most section followed, viz., white-tan, gave the lowest increase, a 2.6 % increase over the XX value as standard. The right-most section, vermilion-forked, gave the highest increase over the control, viz., 15.5 %. The intermediate sections gave intermediate values. That there is a slight increase due to the action of the Y seems probable[1]. That there is a gradient in the action of the Y upon different parts of the X is less probable, but if true, has high interest in connection with our analysis of the mechanism of crossing-over. The modification of crossing-over found in 3N individuals, due to the action of an extra full set of chromosomes, is likewise different in different regions of the X.

---

[1]) New data make it seem probable that tbe percentages of recombination for the regular offspring of XXY females are not different from those of XX sisters.

TABLE 6. — SUMMARY AND COMPARISON OF RECOMBINATION OF SEX-LINKED CHARACTERS IN SISTER XX AND XXY FEMALES

Loci	Tot. (XX)	Rec.	% Rec.	Tot. (XXY)	Rec.	% Rec.	Δ	Δ %
w—t	2660	634	23.8	2043	495	24.2	+ 0.6	+ 2.6
t—v	2660	146	5.5	2043	116	5.7	+ 0.2	+ 3.5
w—v	6165	1859	30.2	6597	2086	31.7	+ 1.5	+ 5.0
w—f	1485	646	43.5	1109	516	46.5	+ 3.0	+ 6.9
v—f	1198	270	22.5	715	186	26.0	+ 3.5	+ 15.5

The striking difference between the XX eggs on the one hand and the X and XY eggs on the other, namely that the X's of the XX eggs are always non-crossovers while the X's of the X and XY eggs are crossovers or non-crossovers in about the normal ratio, was interpreted to mean that the paths followed by these two types of chromosomes had begun to be different at some stage before that at which crossing-over takes place. Such a difference could arise at synapsis, if the X's of the XX eggs followed the type of synapsis that is followed in the male, namely between X and Y. There is no evidence of any crossing-over between Y and X even in the XXY female where such crossing-over might be expected to occur unless the lack of crossing-over were due to a property of the Y itself. If an X were in synapsis with Y, that X would be prevented from undergoing crossing-over. If the other X is left unsynapsed it is also necessarily a non-crossover. When the X and the Y that have synapsed with each other disjoin and pass to opposite poles, this free X should pass to the pole with the disjoined X or to the pole with the disjoined Y. From the first type of reduction come the non-crossover XX eggs; from the second type come XY eggs that are also non-crossovers and that augment the non-crossovers from the normal class of X and XY eggs. If we assume that the free X goes to the same pole with the disjoined X as frequently as to the pole with the disjoined Y, then, to account for the observed frequency of exceptions from XX eggs (4.3%) we must assume that about 16.5 percent of the synapses were of the XY type. If Y were as likely to synapse as X, then 67 % of the synapses should be between Y and one or the other X. The Y seems to be only about one fourth as likely to synapse as compared with the X (BRIDGES '16, p. 18).

A number of YY zygotes, equal to that of the exceptional daughters, should be produced by the fertilization of Y eggs by Y sperm. There is no evidence that these, or the rarer YO zygotes from primary non-disjunction, are ever able to live.

It was long supposed that the XXX zygotes, expected from the fertilization of the XX eggs by X sperm, were unable to live. But recently it has been found that they occasionally appear, but with a mortality above 90 %. They form a distinct somatic type which will be described on page 157 under the heading "superfemale". Here it need only be stated that their finding rounded out the evidence as to the correctness of the explanation of the production of exceptions. The unpublished data of BRIDGES furnish full genetical and cytological evidence of the occurrence and of the constitution of the XXX individuals. They have also been found among the types of offspring produced by 3 N females.

The equality of XX and XXY daughters of XXY females leads to the inference that the regular sons should likewise consist of equal numbers of XY and XYY individuals. The presence of the extra Y would not be detected by any effect upon the visible characters of the male. When fourteen regular sons were out crossed to XX females no exceptions were produced among the F_1 progeny. From each of these outcrosses, daughters were tested to see whether they would produce exceptions. In the case of five sets no exceptions were produced, but in the case of the other nine sets exceptions were produced that were of the frequency of ordinary secondary non-disjunction. The fathers of these nine sets then, were XYY in constitution; and the 5 : 9 ratio is to be regarded as only a chance deviation from equality.

Of 54 daughters of XYY males only 14, or 28 %, failed to give exceptions. If the synapses and reductions in an XYY male take place according to chance there should be a ratio of 33 X-sperm to 67 XY-sperm, which is close to the observed ratio of 28 : 72.

BRIDGES made a cytological investigation of regular sons which showed that both XYY and XY individuals were actually present (unpublished). The previous finding of an XXYY female (BRIDGES '16) had furnished indirect cytological evidence of XYY males.

All exceptional daughters should be XXY, and several that were examined cytologically were all found to be such (unpublished).

These two new items of evidence supplement those previously mentioned, viz., the confirmation of the chromosomal constitutions of regular daughters as XX and XXY, of superfemales as XXX, and of sterile primary males as XO. In aggregate the evidence leaves no escape from the conclusion that those peculiarities of inheritance of sex-linked characters which are grouped under the name non-disjunction are due to the peculiarities of distribution of the rod-shaped chromosome called the X and of the J-shaped chromosome called the Y.

High non-disjunction

The mean percentages of exceptions given by XXY females was 4.3, but occasionally very high percentages were met with. By continuing inbreeding from one of these cultures BRIDGES obtained a line that gave the mean and the mode at 19 percent of exceptions, but there was great variation, the range being from 6 to 36 % of exceptions. Rough selection resulted in several lines that differed from each other considerably in their means, and which had less variability than the original line. None of the cultures gave percentages or lines that were persistently low, so that BRIDGES concluded that there was present a main factor responsible for the high line as a whole and probably several subsidiary factors responsible for the variation and for the different strains. That the modifying factors were not in the X was evident from the fact that the constitution of the X's was constant throughout all the lines. The XXY mothers in each case carried eosin in one X and vermilion and forked in the other, and were mated to Bar males that were brothers. That the main factor or factors were recessive was shown by the fact that F_1 XXY regular daughters from crosses between the high line and unrelated stocks gave ordinary percentages of exceptions. Further evidence of the presence of autosomal modifiers was found in the results given by crossing the exceptional females heterozygous for the high factors to Bar males from the high stock. This was done for six successive generations, during which time the percentage of exceptions rose from ordinary to about 16 %. Then, in the next generation the exceptional daughters were outcrossed to males of unrelated stocks. They themselves gave high percentages, but their exceptional daughters outcrossed for two more generations gave low percentages. During both the rise and the subsequent fall in the percentages the X-chromosome constitution remained constant (One X of the XXY females carried vermilion and forked and came from a high line, the other X carried sable and garnet and came from unrelated stock; the males were first Bar from the high line, and then Bar from another stock).

That sex-linked modifiers were present was shown by the fact that exceptional daughters from outcrosses did not give low but gave about 15 percent of exceptions.

That the two X's of the high stock differed in their effectiveness was shown by the fact that lines derived from equational exceptions usually gave different values than the parent strains, sometimes lower, sometimes far higher than the parent line, depending upon which regions of which X came out homozygous. Some of these strains gave as high as 45% exceptions. Some of the strains in which the X's remained constant and in which rough selection was used gave from 30 to 35 % exceptions.

Since the percentage of exceptions is supposed to be a function of the frequency of heterosynapsis, it seemed likely that an increase in the frequency of the synapsis of Y with X should be produced by a change within the Y itself. However, tests showed that there was no detectable difference between the action of the Y in the high stock and a Y introduced from an unrelated stock.

There are several problems for which high non-disjunction offers better material than does ordinary non-disjunction. For example, as many zygotes are expected to be XXX as are expected to be exceptional females. In cases where this number of exceptions was very high the absence of XXX zygotes was strikingly apparent. Ultimately, however, XXX zygotes were discovered in the high strains, and it was found that certain high strains produced practically none (mortality of XXX 100 %) while other strains produced XXX zygotes whose mortality was roughly 95 % as compared with the exceptional females. In some strains of 100 % non-disjunction (due to the permanent attachment of the two X's to each other, see page 130) the mortality of the XXX zygotes may occasionally become only half that of the exceptional females.

Furthermore, the high percentages of exceptions should be accompanied by a correspondingly high (twice as great) number of absolute non-crossovers added to the normal non-crossover class, provided heterosynapsis is responsible for the fact that the X's of secondary exceptional females are always non-crossovers and provided that the free X passes to the X and to the Y pole equally frequently. That is, high non-disjunction offers opportunity to test these two assumptions. The answer has not yet been obtained, due to the fact that the stocks of high nondisjunction were found to contain modifiers which reduced crossing-over in the X markedly, even after the extra Y was removed. By studying the correlation between the percentages of exceptions and the percentages of crossing-over shown in cultures of a given strain some connection has been established; but the nature of this is as yet too unanalyzed to permit further discussion.

The question as to whether mutation affects both or only one of the two genes originally present in the individual in which the mutative process occurs was long the subject of speculation without means of testing. The work on high nondisjunction furnished two cases in which the mutative process is known to have occurred in only one of the two X-chromosomes without occurring in the other. In the first case an exceptional female carrying eosin in one X and vermilion and forked in the other, gave rise to an exceptional daughter which carried these same chromosomes but in the vermilion forked X a recessive lethal (lethal-8) had appeared. This lethal was located at 59.0 in the X-chromosome, and proved to be a deficiency for the locus of fused. It may be objected against this case that the occurrence of a deficiency is probably not comparable to ordinary mutations; but the second case was that of an ordinary recessive. The recessive sex-linked mutation "bordered" appeared, with its gene in the eosin X of another exceptional female of the same constitution as the above. That this mutation did not affect the other X is certain, since the other X present in the cell in which the mutation occurred was not lost to the polar body but was included with the mutated X in the two-X eggs that produced the exceptional female.

BONNIER, in two papers (BONNIER '22, '23), reports analysis of a strain of high nondisjunction that he received through MOHR. This strain was one of several that had arisen in the work of BRIDGES in which an increase in the percentage of exceptions had followed an equational nondisjunction. The parental culture had given 20.2 percent of exceptions, and the offspring of the equational exception contained 37.6 percent of exceptions. The mean for the succeeding six generations of this stock was 39.5 percent of exceptions. Following this, the stock had been kept for several years in mass cultures without precise knowledge of the percentage of exceptions. BONNIER reports that it gave a mean of about 22 % of exceptions at the time at which he received it through MOHR. Evidently the constitution of the stock, with respect to the autosomal modifiers, had changed during that period of unguarded inbreeding. BONNIER was of the opinion that no autosomal modifiers were present, though his evidence is not adequate on that point. His analysis of the sex-linked modifiers seemed to indicate that they were so numerous that he discarded the idea that they were mutative changes in genes proper, but were rather changes in the "groundwork" of the chromosome, or, as he called it, the "gene-basis". It is by no means certain that these sex-linked modifiers must be assumed to be very numerous, especially in view of the extremely high fluctuating variation present in all lines of high non-disjunction. BONNIER's data show that this unanalyzed fluctuating variability was present in his experiments. His statistical treatment of the data was designed to reduce the probable error by eliminating the extreme fluctuating variants, but in view of the unknown nature of these fluctuations it does not seem to be legitimate to exclude them.

Equational non-disjunction

The ordinary type of primary and secondary non-disjunction gives rise to exceptional daughters whose two X-chromosomes are duplicates of those present in the mother. This result is superficially the same as that which would be brought about by the failure of the reduction division of the X-chromosomes. Hence this type of non-disjunction has been called "reductional non-disjunction". That it is not due simply to a failure of the reduction division of the X-chromosomes is evident from the fact that crossing-over also fails to occur. Very rarely an exceptional female is found whose two X-chromosomes are not duplicates of those of the mother, but instead, one of the maternal X's is represented twice for at least a section of its length. This type of result is similar to that which would be given by a failure of separation of the X's at the equational division, and hence the process, whatever it may be in detail, that gives exceptions characterized by the presence of two representatives of some section of one maternal X, is called equational non-disjunction.

An equational exception is usually detected as an exception that is homozygous for some recessive for which the mother was simply heterozygous, and which the father did not carry at all. Thus, a female carrying eosin in one X and vermilion in the other, and mated to a Bar male, may produce a not-Bar daughter that is homozygous eosin. When this daughter is bred she is found to have one eosin X (a duplicate of one maternal X) and one eosin vermilion X (a crossover X combining the left end of the first maternal X with the right end of the other). Of the nineteen equational exceptions reported by BRIDGES ('16, p. 134) twelve cases had one chromosome that was a non-crossover and the other a crossover. Two others were double crossovers in one X and single crossovers in the other. With respect to the remaining five cases, these also may have involved crossing-over, but that could not have been detected, for the mothers in those cases were heterozygous for only one sex-linked character. Evidently equational non-disjunction is preceded by XX synapsis, and usually involves crossing-over for at least one of the emerging X's. In three cases it is known that the equational non-disjunction occurred in mothers simply XX in constitution. It seems probable that equational exceptions are not due to the presence of the Y; that is, they are always primary exceptions, even when a Y is present.

That the crossing-over evidenced in the equational exceptions did not occur in a two-strand stage is evident; for crossing over in a two-strand stage can give only chromosomes that are duplicates of one another or complements of one another, and can not give two chromosomes one of which is a crossover and the other a non-crossover. The hypothesis that accounts for normal crossing-over and for these exceptions is as follows: XX synapsis occurs; each X splits so that four-strands are present; crossing-over may then take place at a given level between two only of these strands, one from each X (crossing-over between sister strands might occur, but would give only apparent non-crossovers); the reduction division separates the two paternal strands from the two maternal strands (a strand that is a crossover is partly paternal and partly maternal, in this case we may suppose that the whole strand behaves as though paternal if the point of spindle fiber attachment is paternal, and *vice versa*); at its next division the two strands ordinarily enter separate cells, but by an occasional primary non-disjunction the two strands pass to the same pole. One of the two strands could be a single or double crossover and the other a non-crossover; the ratios among the kinds of strands with respect to crossing-over should be the normal ratios. Later evidence from 3N females and from females in which the two X's are attached to one another, likewise can not be accounted for except by assuming that the crossing-over occurred in a stage in which each chromosome was already split into two strands.

The Bearing of Non-Disjunction on Theories of Heredity and Sex.

The above findings from the study of the aberrations in the distribution of sex-linked characters have furnished a positive identification of the rod-shaped chromosome as the carrier of the genes of the whole class of sex-linked characters.

Previous to this demonstration there had been a long series of observations and experiments which had led more and more definitely to the chromosomes as the bearers of the hereditary materials. It was observed that male and female contribute equally to the inheritance of the offspring and yet the contribution of the male consists of little more than a nucleus. That inheritance is a function of the nucleus rather than of the cytoplasm was shown by many embryological and cytological facts. Attention was next narrowed to the chromatin and chromosomes. Embryological experiments showed that the chromosomes are qualitatively different, and that at least one of each kind is essential for normal development. The increasing cytological knowledge of mitotic division and of gametogenesis made it clear that the chromosomes are qualified to serve as the material basis of heredity. The next advance was the result of the exact knowledge of heredity which MENDELIAN analysis furnished, and the coupling of experimental genetics with cytological investigation. It was shown that the genes for character differences and the chromosomes have the same method of distribution, as exemplified in segregation and in free assortment. Sex-linked inheritance and the sex-chromosomes furnished cases in which genes and chromosomes had the same normal distribution. The final step was the demonstration of the identity of distribution between specific genes and a specific chromosome, in such a way that the argument of "the cell-as-a-whole" cannot be applied, and in such a way that the chromosomes must be regarded as the means and not the consequence of the inheritance of characters and sex. The distribution of sex-linked genes (as followed by experimental breeding methods) has been demonstrated to be identical, through all the details of the normal process and of a highly complicated new situation, with the distribution of the rod-shaped chromosome (as tested by direct cytological examination). The argument that the cell-as-a-whole possesses the tendency to develop a given Mendelian character is completely nullified by the fact that in these cases the cells that produce a given character (exceptions) are of exactly the same origin as those which produce the allelomorphic character (regulars), the only difference being confined to the fate of a particular chromosome, the rod-like "X-chromosome". Those eggs that have lost nothing but the X-chromosome have completely lost therewith the ability to produce any maternal sex-linked character, and with the introduction of an X from the father those eggs have developed all of the sex-linked characters of the father. Conversely, those eggs that have retained both X's of the mother and have received no X from the father show all the sex-linked characters of the mother and none from the father.

Likewise experimental proof is given that a particular chromosome, the rod-like "X", is the differentiator of sex; the chromosome constitution of an individual is the cause of the development by that individual of a particular sex, and is not the result of sex already determined by some other agency. Sex is not determined in the egg or sperm as such, but is determined at the moment of fertilization; for, the X-sperm of a male (that ordinarily gives rise to a female) gives rise to a male when it fertilizes an egg having no X-chromosomes. Likewise the Y-sperm of a male (that ordinarily gives rise to a male) gives rise to a female if it fertilizes an egg containing two X-chromosomes. Such facts prove that the segregation of the X-chromosome is the segregation of the sex-differentiator. The presence in the zygote of two X-chromosomes determines that the individual shall be a female, the presence of only one, that the individual shall be a male. The origin of these chromosomes, whether maternal or paternal, is without significance in the production of sex. The role of the autosomes in the production of sex will be discussed in the section on intersexes and supersexes.

CHAPTER XII

THE ATTACHMENT OF X-CHROMOSOMES TO EACH OTHER

A fly was found by L. V. MORGAN ('22) all of the offspring of which were non-disjunctional exceptions, due to the fact that the two X-chromosomes of the mother were permanently attached to each other, or to the same spindle-fiber.

In a mating between a not-yellow female and a yellow male a mosaic occurred in which the head and thorax were, as expected, not-yellow, but in which the abdomen was yellow like the father. This yellow abdomen was entirely female; and the mosaic bred as a female when crossed to a not-yellow male. All of the daughters of this cross were yellow like the mother, and all of the sons were not-yellow like the father! That is, there was a complete reversal of the ordinary "criss-cross" inheritance in which a yellow mother mated to a not-yellow father gives only not-yellow daughters and yellow sons. All of the offspring were exceptions; and the hypothesis was advanced that this 100 % non-disjunction was due to some sort of union of two X-chromosomes, each bearing the yellow gene. This union, or attachment, was supposed to have occurred in some cell of the germ track of the yellow father of the mosaic, and a sperm carrying these attached X's had fertilized an egg carrying a not-yellow X. The resulting zygote was a triplo-X, a type of fly which usually dies, but which in this case was able to live more readily since one of the X's (the maternal not-yellow X) was eliminated from some of the cells, leaving a large part of the fly (the yellow abdomen) with only two X's. The attached X's remain united permanently; and hence the mature eggs contain both or neither, these two types being equally frequent. The double-X egg (yellow) by Y-sperm gave yellow daughters, while the no-X eggs by X-sperm gave the not-yellow sons (Fig. 45).

Three further genetical facts strengthened the hypothesis above outlined. First, all of the not-yellow sons of the above mosaic were found to be sterile. This confirmed the expectation that they were XO in constitution, and came from eggs without a sex-chromosome fertilized by X-sperm. Second, the yellow daughters of the mosaic, outcross-

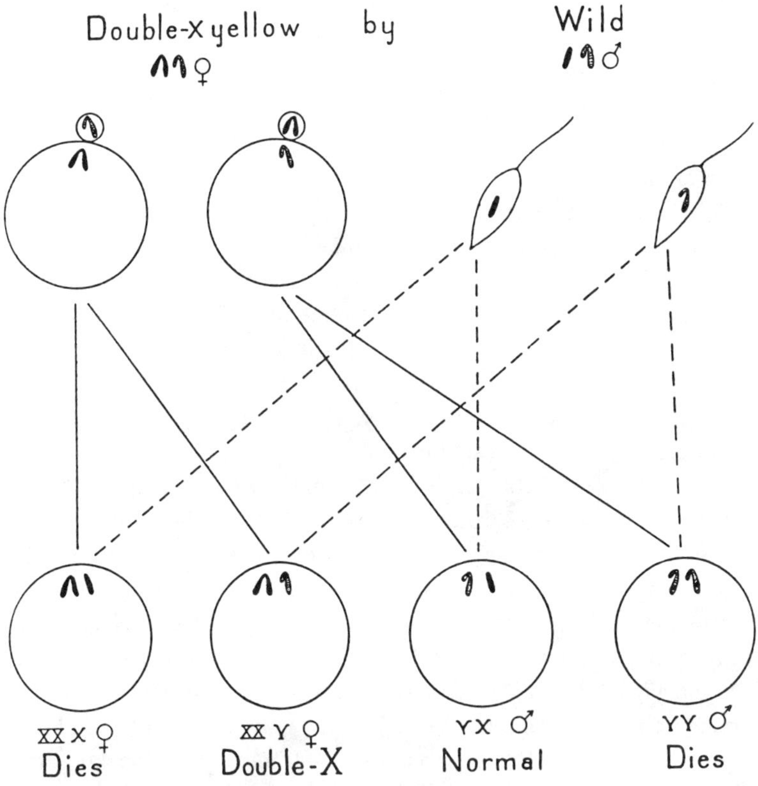

FIG. 45. Complete non-disjunction, due to attached X's. The double- X is a large V-shaped chromosome, both components of which carry the recessive gene yellow. A Y-chromosome (stippled) is present with the double-X.

ed to males carrying any sex-linked characters, gave again only yellow daughters, and sons like the father in each case. These sons were always fertile, which showed that each mother had carried a Y-chromosome, and had come from a double-X egg fertilized by a Y-sperm. Third, in subsequent generations cultures were found in which considerable numbers of the triplo-X superfemales, expected from the fertilization of the double-X type of egg by the X type of sperm, were present and able to emerge.

The proof that the hypothesis was correct came from a cytological examination of the exceptional yellow daughters and of the superfemales. The yellow daughters were all found to have a Y-chromosome and two X-chromosomes that were attached, at their inner ends only, to form a V-shaped chromosome similar to, but larger than, the V-shaped autosomes, which were present in normal condition (Fig. 10). The figures of the superfemales were like those found by BRIDGES for regular superfemales (Fig. 57), except that two of the three X's were attached to each other and appeared as a large V-shaped chromosome in company with the rod-like third X.

The double-X race is very valuable in securing and maintaining a stock of males with one or more sex-linked characters. Thus, any desired combination, when secured as a single male, can be reproduced in any desired quantity by mating that male to double-X females. Also, such a stock maintains itself automatically without selection.

Occasionally the two X's of the double-X break apart, giving a single-X egg, and hence a yellow son or a not-yellow daughter. This event is too rare to disturb the usefulness of the stock.

A second case of attached X's was found by E. G. ANDERSON, who furnishes us with the following hitherto unpublished facts with respect to the case. In order to secure a large number of equational exceptions for the purpose of studying crossing-over and reduction, ANDERSON had X-rayed females that had carried the recessives scute, apricot, cross-veinless, tan and forked in one X, and in the other X, broad, echinus, cut, and garnet. One of the equational exceptions had showed the characters broad and echinus. When she was mated to a not-broad not-echinus Bar male, all the sons were like the father and all the daughters were broad and echinus; that is, she gave complete non-disjunction. The exceptional broad echinus daughters, in turn, have given complete non-disjunction in each generation. The results can be interpreted by the assumption that in some cell in the germ track of the X-rayed female, the two X-chromosomes became attached to each other, giving a double-X of the type previously found by L. V. MORGAN. That the exceptional daughters in this line also, do carry such a double-X was verified by cytological examination.

The original equational exception was broad echinus, that is, a fly homozygous for two characters for which the mother had been simply heterozygous. Moreover, some of the exceptional daughters in each subsequent generation showed also other characters for which the broad echinus mother had been only heterozygous. The broad echinus mother was found to be of the following constitution:

```
br      ec      cv       t       f
_____________________________________
_____________________________________
br      ec               ct      g
```

In order that such a mother may give rise to an exceptional daughter homozygous for cut, in addition to broad and echinus, it is necessary that two X's emerge each of which carries the cut gene. When such a br ec ct female is bred, she also is found to give complete non-disjunction, and among her exceptional br ec ct daughters some are homozygous for one or more of the other recessives for which the br ec ct mother was heterozygous. That is, the two X's of which she is composed have become alike for part of their length without becoming alike for their entire length. The daughter X's in these cases are neither identical nor complementary, and hence these cases add further proof to that adduced from the equational exceptions (discussed on page 127), and the findings with respect to the offspring of 3N females (see page 161), that crossing-over occurs at a stage at which four, and not two, strands are present.

A curious relation was discovered with respect to the frequency with which the various heterozygous mutants emerge as homozygotes. This frequency is greatest for crossveinless, cut and tan, markedly less for garnet, and least for forked.

That is, the further to the right the locus, the greater is the difficulty in becoming homozygous. For a heterozygous recessive to emerge as a homozygote in attached X's requires crossing-over between the point of attachment and the locus of the recessive; the farther away the locus the more frequent is the emergence of the recessive. From this relation ANDERSON concludes that the point of attachment of the two X's is to the right, beyond forked, and the free ends are the left ends. In other words, the spindle-fiber attachment to the X-chromosome is at the extreme right end (presumably beyond the locus of bobbed) and not (as previously supposed to be the case) at the left end where yellow is located.

A third case of attached X's was found by A. H. STURTEVANT (also unpublished) shortly after the case of ANDERSON. Here also cytological proof that the X's are attached has been obtained. The two X's in this case were both derived from a wild-type male, and were therefore alike. This would happen in a male if at an ordinary division the two daughter halves of the X-chromosome remained united.

But triploidy arose in this strain, and it was found that, in 3N females with two of the three X's attached, crossing-over occurs between the free X and one of the attached X's. Because of this fact it is possible to obtain attached X's carrying any desired combination of genes, either in homozygous or in heterozygous form. The behavior of attached X's in this strain has been studied chiefly in 2N females, derived from 3N ancestors, that were heterozygous for sc, ec, cv, ct, v, g, and f. Such females give about 1/6 of their daughters homozygous for any one of the recessive genes from sc to v, inclusive, rather less than $^1/_6$ homozygous for g, and about 6 % homozygous for f. A small series heterozygous for f, B, and fu has also shown that the attachment point must lie to the right of fused. A detailed study of the crossing-over in attached-X females is now being carried out on this line.

In the original yellow double-X strain of L. V. MORGAN triploidy arose also; and from this 3N strain 2N females were derived in which the attached-X's carried various recessives. The frequencies of appearance as homozygotes of the mutant genes that are present in only one of the attached X's, show that in this strain also the spindle-fiber attachment is at the "right" end.

These cases furnish a new general method for interpreting the alteration of chromosome number in related forms. The changing of the chromosome number by the compounding of non-homologous chromosomes is familiar in the species in which the X is attached to an autosome, or in which two autosomes, a^1 and a^2, have been compounded to form a single larger a^1a^2-chromosome. In this new method the compounding is between homologues, and a double-a chromosome is produced. Such double-chromosome forms could breed true, with the elimination in each generation of half the zygotes, viz., the nullo-a types and the tetra-a types. Some tetra-a types may be viable and give distinct forms, homozygous double-a forms. These would differ cytologically from the parent form in having, instead of a pair of rods, a pair of double-size chromosomes, usually V-shaped.

CHAPTER XIII

Non-disjunction of the Fourth Chromosome

For several years after non-disjunction of the X-chromosome had been discovered a close watch was kept for phenomena that would indicate non-disjunction of an autosome. Several peculiar cases of inheritance were found that at first were taken to be such, but further investigation showed that they were due to other types of disturbance. That the autosomes were not subject to occasional non-disjunction did not seem probable, and accordingly it was supposed that the zygotes produced as a result of non-disjunction (the 2N + 1a and 2N — 1a types) were so greatly changed somatically as to be lethal. This still seems to be true for non-disjunction of the large V-shaped autosomes, chromosomes-II and -III, but in the case of the small round chromosome, the zygotes (the triplo-IV and haplo-IV types) are now known to occur and to be fairly viable.

The haplo-IV Type

A rather frequent type of mutant character in *Drosophila melanogaster* is that called "Minute", which is characterized by small, slender bristles, rather large, roughish eyes, blunt wings, and by several minor differences in wings and coloration (Fig. 46).

The Minutes are dominants, that are lethal when homozygous, and generally are much delayed in hatching. Among the Minutes one particular type came to be recognized as of frequent recurrence and characteristic genetical behavior. This distinct type was called "Diminished". The breeding results showed that Diminished behaved as a "deficiency" (see page 163) for all the fourth-group mutants. Among the offspring of Diminished flies exceptions to the normal distribution of fourth-group characters occurred that showed that the Diminished flies were produced by non-disjunction of the chromosome carrying the fourth group, and were zygotes that lack one member of the fourth-chromosome pair. The deficiency phenomena in this case were due to

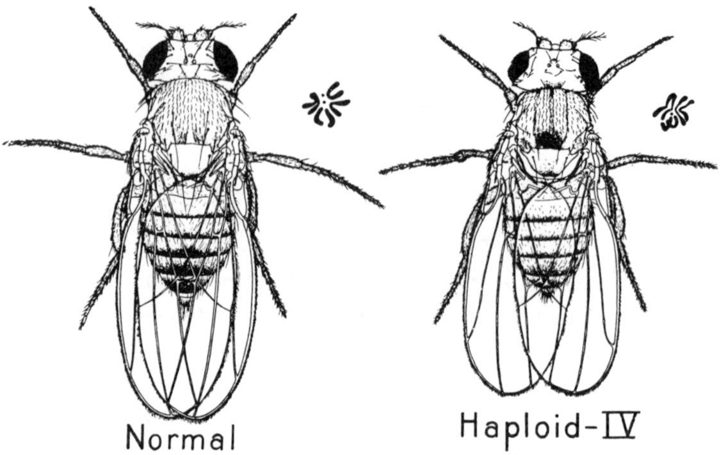

FIG. 46. Haplo-IV female with a chromosome group, compared with a normal female and its chromosome group.

the loss of an entire chromosome. Cytological examination of the cells of Diminished individuals showed that one of the small round chromosomes was absent. This finding identified the small round chromosome as the carrier of the fourth group of linked genes.

The cytological proof that the characteristics of Diminished are due to the loss of a fourth chromosome has put the theory of deficiency on a sound basis; for, each of the features met with in the cases (Bar-deficiency, BRIDGES '17, vermilion-deficiency, BRIDGES '19, the Notch-deficiencies, MOHR '22, etc.) previously interpreted, from genetical data, as due to the loss of a section of chromosome, has also been met with in the case of Diminished; and it has thus been shown, not only that these features *might* be produced by loss of chromosomal material, but that in this specific case they *are* actually produced by such a loss.

The general interpretation of the effects of deficiency rests on the theory that each character of the wild-type is due to the united action of many genes, some of which tend to bring about development in one direction and others in the opposite direction. The point of balance between the set of "plus" and the set of "minus" modifiers of each character determines the grade of development that is shown in the wild-type (see BRIDGES '22, for a full presentation of the conception of "genic balance"). These "plus and minus modifiers" are apparently distributed at random along the chromosomes, so that if a particular short section is removed, as in a sectional-deficiency, or an entire small chromosome, usually it will happen that more plus than minus modifiers (or vice versa) for one or for several characters, will be removed. Consequently, relatively more minus than plus modifiers (or vice versa) will be left in action, and a mutant character, or complex of characters, will appear, differing from the old in a minus direction (or vice versa).

The features of Diminished that are characteristic of deficiency are, briefly, as follows: —

Diminished flies are characterized by a complex of mutant characters, some of which are plus and others minus departures from the wild type (see Fig. 46). Thus, the body-size is smaller throughout; the bristles are relatively shorter and more slender; the body-color is paler; the thorax-pattern is darker; the eyes are larger and rougher, but with smaller hairs; the aristae are reduced or absent; the wings are slightly spread apart, are blunter, and cloudier in texture. To these plainly observable somatic alterations may be added certain other characteristics that are interpreted as consequent upon changes in internal structures, viz., low productivity, frequent sterility, slow development, and heavy mortality. Diminished individuals begin to emerge four or five days later than wild-type sibs. In crosses in which half of the offspring were expected to be Diminished a mean of only 34.5 per cent was found; that is, the viability of Diminished was about 53 per cent as compared with the wild-type. The mortality of Diminished was found to be very variable, and the cultures were apparently of two types, one mode being at about 10 per cent of Diminished, and the other at about 35 per cent. No analysis of these differences in mortality was attempted. The high mortality of Diminished individuals mainly depends upon changes during the pupa stage. For, if the mortality were largely due to changes in the larval stage, then mass-cultures, because of the greater competition among the too numerous larvae, should show a lower proportion of Diminished than do sister pair cultures, but this was found not to be the case.

All of the above characters appear in the F_1 Diminished that come from outcrosses, that is, they are dominant. When Diminished flies were bred together a mean ratio of 139 Diminished to 100 wild-type was obtained. This, and other evidence, showed that "homozygous Diminished" (that is, the nullo-IV) is lethal. The lethal effect of homozygous deficiencies in general is accounted for by the total loss of both representatives of the section of genes, which may be expected to produce far greater changes in the balance of the genes than does the removal of only one of the two homologous sections.

The constituent characters of the type Diminished always appear together, which means that no crossing-over has been occurring throughout the deficient section.

The type Diminished was found not to be sex-linked and not to show linkage with second or with third-chromosome mutants. Since there are only four linkage groups, the above evidence showed that Diminished is produced by a change in the fourth group of linked genes. Back-crosses of Diminished with previously known fourth-group mutants should be found to give strong linkage. But when Diminished was crossed to bent, to eyeless and to shaven, to secure the necessary double heterozygotes, it was found that all F_1 flies that were Diminished were at the same time bent or eyeless or shaven, respectively. Such "pseudo-dominance" is expected when dealing with deficiencies; for, the situation is then similar to the case of the appearance of sex-linked recessives in the male. The Diminished individuals are haploid for all IV-chromosome loci, including the loci of bent, eyeless, and shaven, and when the IVth chromosome that is present carries the gene for one of those characters there is no wild-type allelomorph present to dominate over the effect of the one recessive gene.

The characters bent, eyeless and shaven, as they appeared in the Diminished flies, were more extreme departures from the wild-type than are the simple characters of the respective mutants. This effect of a deficiency upon characters is called "exaggeration", and is interpreted as due to a different ratio, within the region involved in the deficiency, of plus and minus modifiers with respect to the mutant character considered, than is present in the normal 2N situation. The same thing is observable in the case of sex-linked mutant characters that are different in the male (deficient for one X-chromosome) than in the female. Most of the "sex-limited" sex-linked characters are exaggerated in a plus direction in the male; that is, they are more extreme departures from the wild-type, but some are "exaggerated" in a minus direction.

The mortality of the bent, eyeless, and shaven mutants in Diminished flies is also exaggerated. The mortality of eyeless, of bent, and of shaven is roughly 5, 10, and 20% above that of the wild-type, while the mortality of Diminished is roughly 50% above standard. But the mortality of eyeless in Diminished was exaggerated to include 98.7, of the zygotes, of bent to 95.5 and of shaven to 100 %. This exaggeration of mortality means that not only the superficial characters are exaggerated, but also the physiological changes or changes in internal organs. The mortality of the compounds of Diminished with eyeless, bent, and shaven, as well as the mortality of simple Diminished, is mainly due to changes in the pupa stage. For, if a culture bottle, in which the offspring of a Diminished by eyeless mating have been reared, is examined, it is found that the remainder of the 50 % of Diminished-eyeless individuals that were expected, are present as dead pupae.

Certain characters not in the fourth linkage-group, notably Hairless, are exaggerated in Diminished. This is parallel to the exaggeration of autosomal mutants by deficiency for the X-chromosome, that is, to the whole class of sex-limited autosomal characters.

The genetical features of Diminished that have been given in the above sections do not require that the deficiency be assumed to cover the entire fourth chromosome — a sectional-deficiency covering the section with the loci for bent, shaven, and eyeless would suffice — but there are two genetical features incompatible with the sectional-deficiency view, viz., a very high frequency of recurrence, and the production of exceptions to normal inheritance. The type Diminished has been recorded some twenty-five times within a year. Notch-deficiency has recurred some twenty-five times in ten years; but there is good evidence that the various Notch-deficiences differ in the length of the section lost and also, therefore, in the various characteristics, while the various recurrences of Diminished have given no such indication of heterogeneity. As a sample of the exceptions to normal inheritance may be cited the appearance, in a cross of Diminished to bent, of a Diminished that was not bent, and that did not carry bent although the father had been homozygous for bent. This was interpreted to mean that a normal haplo-IV egg of the Diminished parent had been fertilized by a non-disjunctionally produced nullo-IV sperm of the bent father. A second case arose in the appearance, (in a cross of Diminished to eyeless) of a not-Diminished eyeless individual, which may be explained as due to a normal nullo-IV egg of Diminished, fertilized by a non-disjunctionally produced diplo-IV sperm of the eyeless father. That the deficiency phenomena and the exceptions were alike due to loss of one of the fourth chromosomes of Diminished was proved by direct cytological examination of some seventy-five Diminished females in which satisfactory figures were found.

The haploid nature of the Diminished individuals made possible a back-cross determination of the bent-eyeless recombination per cent, without first obtaining the bent eyeless double recessive. The situation in the fourth chromosome is now comparable to that in the first, with haploidy for the fourth chromosome giving the Diminished type and pseudo-dominance of fourth-group recessives, while haploidy for the first chromosome gives the male type and pseudo-dominance of first-group recessives. Females heterozygous for bent and for eyeless were crossed to Diminished males, and the offspring furnished the equivalent of 17 recombinations in a total of 1787, or 0.86 ± 0.14 percent of recombination.

The positive identification of the small round chromosome as the carrier of the genes of the fourth-group mutants, extends to autosomal characters the proof of the chromosome theory of heredity.

The triplo-IV

It is evident that the same non-disjunctional process that gives rise to nullo-IV gametes and to Diminished individuals (haplo-IV's) should as frequently give rise to diplo-IV gametes and hence to triplo-IV zygotes. The frequency of the production of haplo-IV gametes is roughly one in 2000. In spite of a close watch no individual was found whose somatic appearance or breeding peculiarities agreed with those expected for a triplo-IV individual.

Such individuals (Fig. 12b) were ultimately found among the offspring of 3N females. In a 3N female, a large proportion of the gametes have one X and one set of autosomes (see page 156), but since the 3N mother has three IV-chromosomes, this general class of 1N gametes should include, besides those strictly 1N, an approximately equal number of gametes having an extra fourth chromosome (N + IV). That is, among the offspring classed as females and males there should occur an admixture of triplo-IV individuals. When such a possibility was recognized, a search was made among the offspring of 3N females for triplo-IV individuals. None were distinguished from the regular wild type, which showed either that they were not present or that they were so closely similar to the wild-type that special methods would have to be used in identifying them. By this time the idea of genic balance had been developed; and accordingly it was expected, since the loss of one fourth chromosome gave rise to a specific complex of mutant characters, some of which were minus and others plus deviations from the wild-type, that the gain of one fourth chromosome would give rise to a corresponding complex of mutant characters, but each a deviation in the opposite direction. That is, the next step was a deliberate search for individuals that should have relatively narrow, pointed wings, of clear texture, and held close together; whose eyes should be small and smooth; whose general body-color should be dark; and yet whose trident pattern should be less strikingly developed; with relatively coarser bristles; etc. Presently a class of flies was distinguished that possessed these characters, however, in slight degree. The clearest characters were the small, smooth eyes, the narrow, more pointed wings, the darker body-color, and the suppressed trident pattern.

This type of flies was isolated and bred, and the complex of slight differences was found to be inherited as a unit and as a dominant, as it should be if due to the presence of an extra fourth chromosome.

The "triplo-IV" type was tested genetically in a variety of ways, the most significant of which was by means of a cross to a fourth-group mutant, (Fig. 47) e. g. eyeless. From the F_1 of "triplo-IV" to eyeless, "triplo-IV" flies were selected and back-crossed to eyeless. The ratio of not-eyeless to eyeless in those results, instead of being the usual 1 : 1, was approximately 5 : 1. As seen in the diagram of Fig. 47, this finding agrees with expectation, if it be assumed that in a triplo-IV individual one wild-type gene is dominant to two eyeless genes. When F_1 "triplo-IV" flies were bred to each other the ratio of not-eyeless to eyeless was approximately 26 : 1, and not 3 : 1. Also, the fact that the ratio was found to be approximately 26 : 1 and not approximately 36 : 1 is interpreted as due to the death of the tetra-IV type. The death of the tetra-IV type is also indicated by the fact that when triplo-IV were inbred, no type was found that conformed to expectation for a tetra-IV somatic type, viz., a complex of characters similar to that of haplo-

IV, but with opposite signs, and similar to and with the same signs as that of the triplo-IV, but whose degree of departure from the wild-type is greater than that in triplo-IV. Evidently the degree of depar-

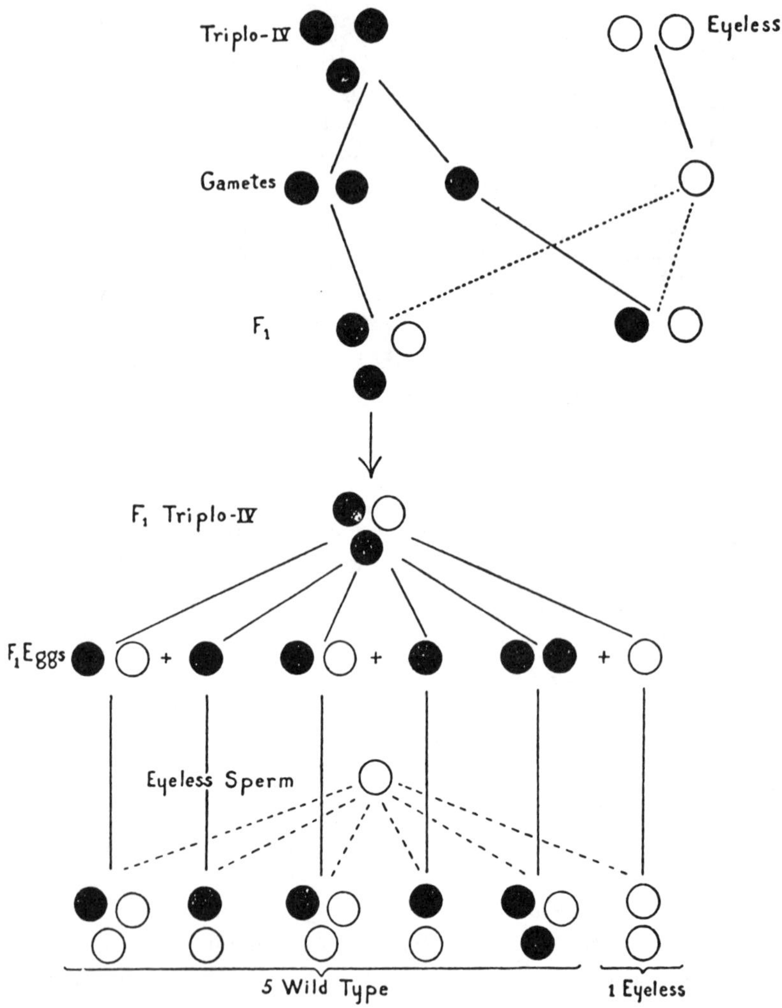

FIG. 47. The result of mating a triplo-IV to a IV-chromosome recessive, eyeless, and back-crossing the F_1 triplo-IV to an eyeless male.

ture in the tetra-IV is so great, with respect to some characters, that the zygote is inviable.

That the supposed triplo-IV type is actually such, was proved by cytological examination of the germ cells of females that were classified as belonging to this type. Good figures were found in about ten such females; and in each case it was found that there was present an extra, small, round chromosome (see Fig. 12b).

CHAPTER XIV

Gynandromorphs and mosaics

Elimination of the X-chromosome

In the course of the Drosophila breeding-work some two or three hundred flies have been found which were mosaics with regard to sex; that is, they were "gynandromorphs" (for full account see Morgan & Bridges '19). Most of these were bilateral gynandromorphs, that is, individuals that were female on one side and male on the other, with a sharp line of juncture of the two kinds of tissue. Since the male is normally smaller throughout than the female the body of the gynandromorph is usually warped toward the male side (as seen in Fig. 36G, in the Mechanism of Mendelian Heredity 2nd ed., p. 118. Fig. c represents one of the first gynandromorphs found.) The hairs and bristles are also smaller on the male side; and the line of juncture can be traced rather accurately by this means, as well as by the smaller male eye, the smaller male wing, the male sex-comb on the front leg, and by the darker coloration of the male regions in the abdomen. All these features show in the specimen drawn in Fig. 49. Much more frequent than the strictly bilateral gynandromorphs are those in which female

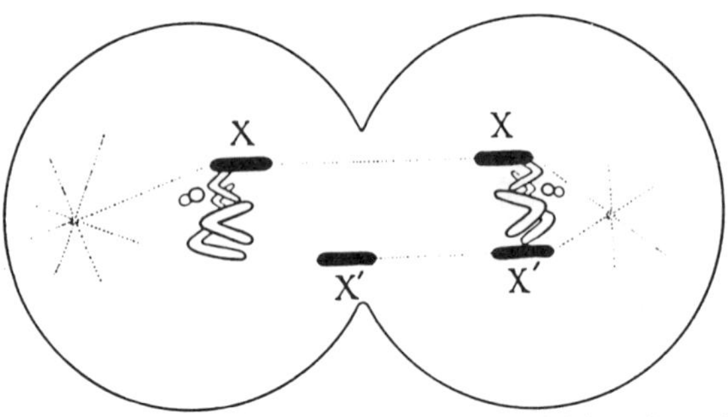

Fig. 48. Elimination at an early cleavage division of one of the daughter X-chromosomes of the paternal X (X'). The parts descended from the cell at the left will be male and maternal, the parts descended from the cell on the right will be female and hybrid in type.

parts predominate; thus, the most frequent type is one that is bilateral except that the head is entirely female, or is bilateral except that the abdomen is predominantly female. In some cases a quarter only is male, or even a still smaller region. More rarely the gynandromorphs are "fore-and-aft", e. g., with a female head but male body and abdomen.

The gynandromorph drawn in Fig. 49 is bilateral throughout, the left side of the head, thorax, and abdomen being female and the right

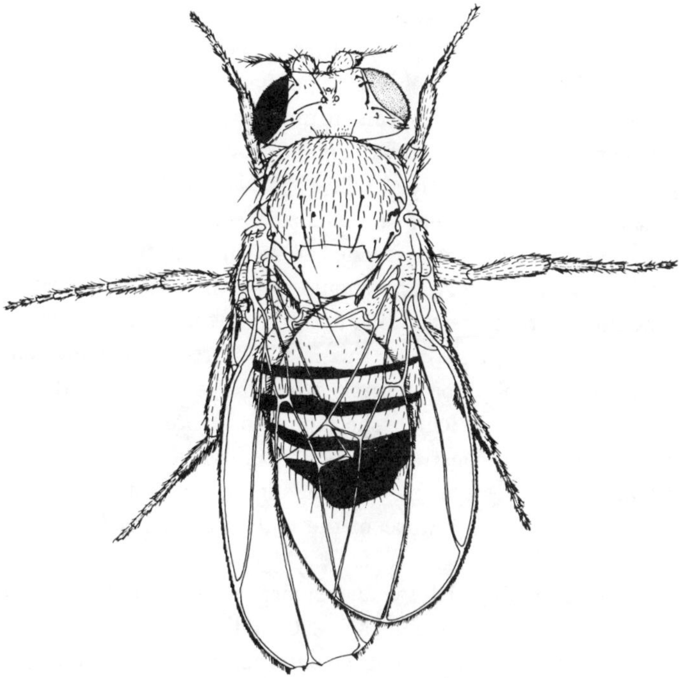

FIG. 49. A gynandromorph whose left side is female throughout, and shows the dominant character Notch, and whose right side is male throughout and shows scute broad echinus ruby tan and forked.

side male. The left side shows the dominant character Notch (its mother was Notch) and right side shows the sex linked recessive characters, scute, broad wing, echinus eye, ruby eye, tan body color and antenna color and forked bristles (all of which entered from the father).

Whenever as here, there are sex-linked characters in the cross that produces the gynandromorph, the sex-linked characters of the male parts correspond either to those of the mother or to those of the father, while the characters of the female parts are those of the hybrid between the maternal and paternal combinations. On the other hand, autosomal characters appear in the two sides without distinction. This suggests that the cause of the gynandromorph is to be sought in an aberration of the X-chromosome alone. It is supposed that this type of gynandromorph starts as a female, and that the male parts arise through loss of one of the two X-chromosomes. For example, the gynandromorph of Fig. 87B page 190 "Physical Basis of Heredity" revised ed. 1923, had for its parents a wild-type female and an eosin miniature male. The female parts are those expected to be present in the hybrid from the cross, viz., the wild-type characters red-eyes and long-wings that are dominant to eosin and to miniature. But at an early cleavage division, probably the first (Fig. 48), one of the two daughter X's of this maternally derived wild-type X was lost, "eliminated". This left a cell that had only one X-chromosome, the paternal eosin-miniature X. All the parts descended from this cell should be male, since they have but one X, and should show the paternal characters eosin and miniature, corresponding to the X that was retained.

There is other evidence that this is the correct interpretation of these gynandromorphs; for example, when the genitalia and abdomen of a gynandromorph are female she can often be bred, and then she is found to be of hybrid constitution. Moreover, when the abdomen and genitalia are male, the gynandromorph is found to be sterile, which is expected from the XO nature of the male cells in these gynandromorphs.

The frequency with which gynandromorphs are found in the Drosophila material is roughly one in each 2000 individuals.

The frequency of elimination of the paternally derived X is the same as that for the maternally derived X; for, the male parts show sex-linked characters derived from the mother as frequently as from the father.

If the elimination occurs at the first cleavage division the gynandromorph is expected to be bilateral except for deviations due to subsequent wandering of the cells. That such wandering occurs is evidenced by the large number of cases in which the line, especially in the head, is shifted to one side or takes an irregular course.

In bilateral gynandromorphs the two gonads have nearly always been found to be the same, i. e. both are testes or both are ovaries; and it was formerly thought that this was a constant feature. However, exceptions have occurred, and the cytological work of HUETTNER ('22) has shown that several nuclei of the embryo give rise to the primordial

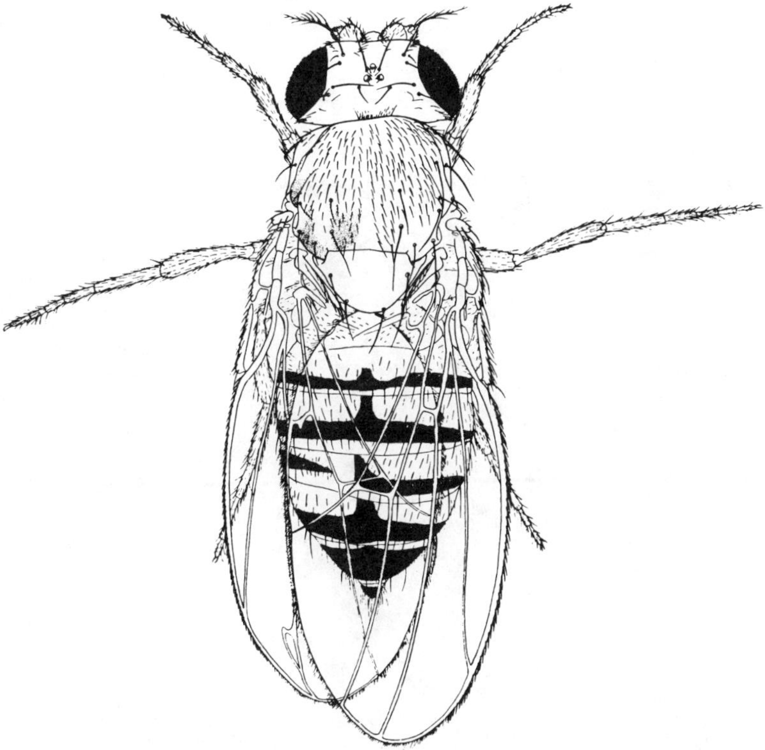

FIG. 50. A mosaic by elimination of one IVth chromosome, the left side being haplo-IV and the right side normal diplo-IV.

germ cells, hence the gonads are not as formerly supposed, derived from a single nucleus lying at the posterior end of the egg.

A striking fact with regard to these gynandromorphs is that the male and female parts and their sex-linked characters are strictly self-determining [1]), each part developing according to its own genetic constitution. No matter how large or small the region may be, it is not interferred with by influences coming from neighboring parts, nor is it overruled by the action of the gonads.

---

[1]) Except in the case of vermilion eye (Sturtevant, '20b).

Elimination of the fourth-chromosome

Parallel to the cases of mosaics for sex, and for sex-linked characters due to elimination of an X-chromosome, are mosaics for Diminished and for IV-chromosome characters, due to elimination of the IV-chromosome. (Bridges, unpublished).

In crosses free from Diminished, probably more than a dozen flies,

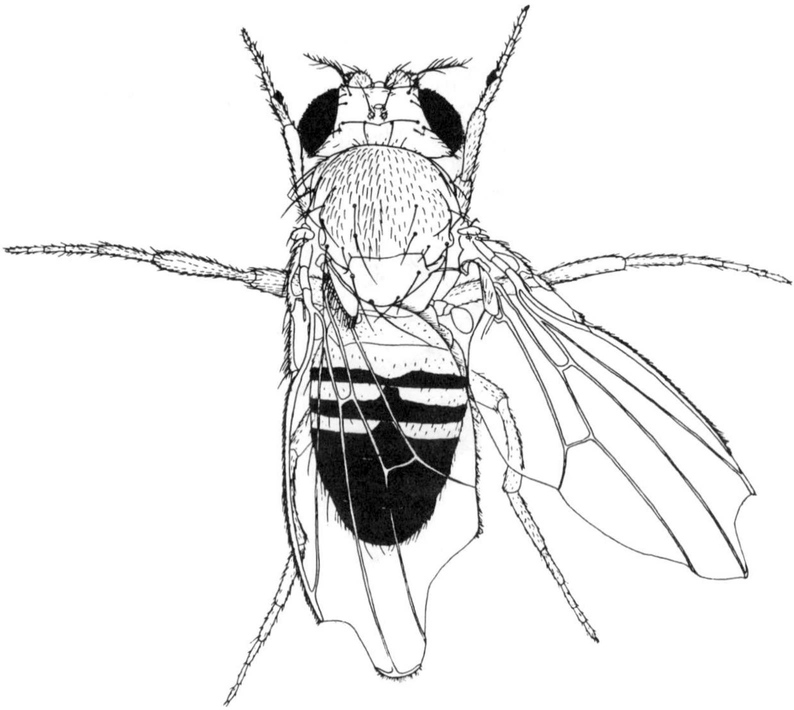

FIG. 51. A male (cut-12) that is haplo-IV on the right side and normal on the left.

some male, some female, have been found in which the body was sharply divided into a normal part and a part showing the characteristics of Diminished (Figs. 50, 51). Some of those may be really somatic mutations to a Minute, but certainly the majority must be due to elimination of one of the two fourth-chromosomes at an early cleavage division. In a cross of eyeless by wild-type, a fly occurred (Fig. 52) that was wild-type on the left side and apparently eyeless-Diminished on the right. If the classification was correct, this fly can be interpreted as one that started as a normal heterozygote for eyeless, and in which

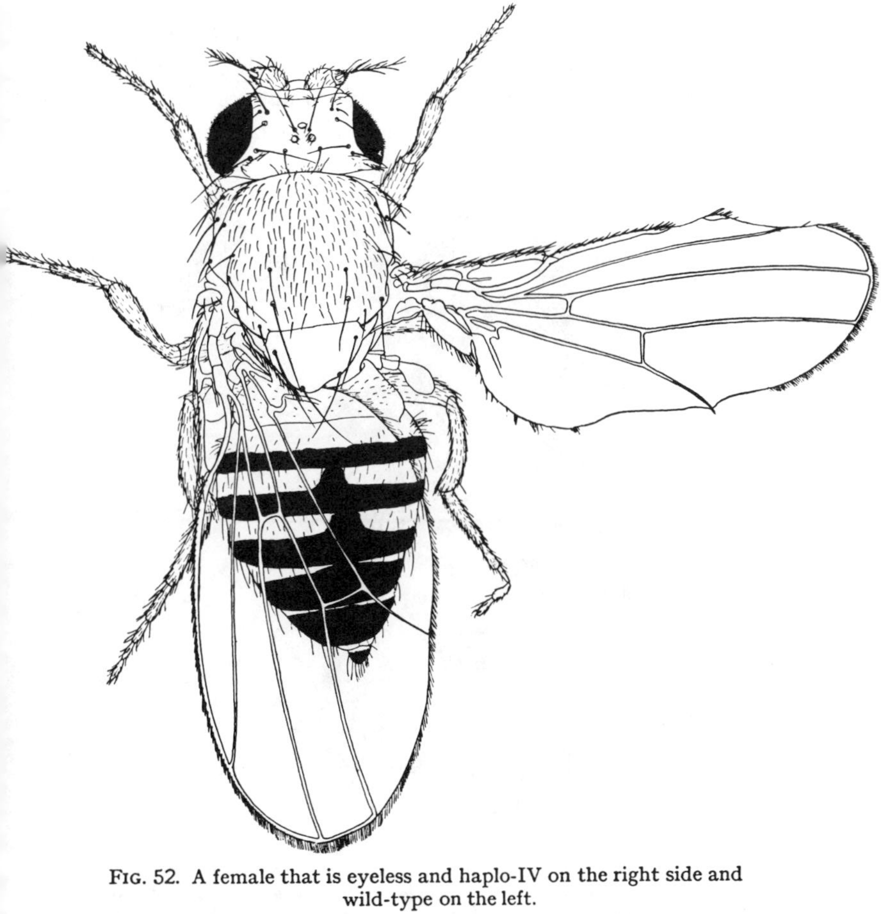

FIG. 52. A female that is eyeless and haplo-IV on the right side and wild-type on the left.

elimination of the not-eyeless chromosome occurred at an early cleavage division.

The double-nucleus gynandromorphs and mosaics

Besides the very frequent gynandromorphs due to elimination of one of the two X-chromosomes in part of an embryo that started as a female, or to elimination of one fourth-chromosome in part of a diplo-IV embryo, there are about twenty gynandromorphs and mosaics, including mosaics for autosomal characters, which require, or are better fitted by, the hypothesis that they have arisen from eggs that contain two seperate female pronuclei, each of which is fertilized by a separate sperm. The evidence from the offspring of 3N females makes it certain that zygotes having an extra second or third autosome are inviable, and other evidence makes it certain that zygotes lacking one such au-

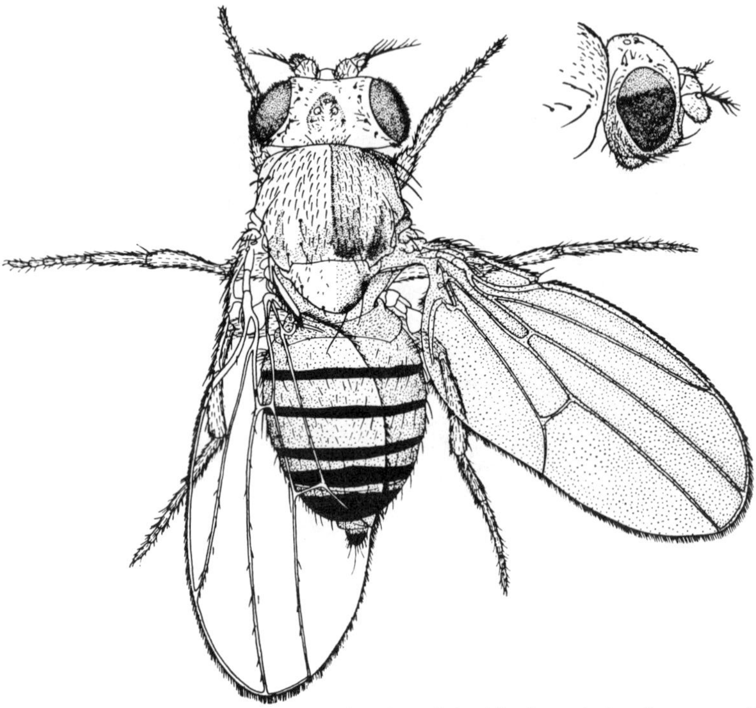

FIG. 53. A double-nucleus mosaic whose left side showed the three recessive third-chromosome characters hairy, scarlet and spineless, and whose right side showed the dominant Dichaete and the recessives peach and sooty. The division line passed through the right eye, which was scarlet above and peach below.

tosome are lethal. Hence, when mosaics are found that involve second or third-chromosomal characters some other explanation than autosomal elimination or non-disjunctional aberrations in distribution must be sought. A striking case in which several third-chromosome characters were involved was found by BRIDGES. A female that carried the four mutant genes roughoid, Dichaete, peach and sooty in one third-chromosome, and in the other the three recessives hairy, scarlet and spineless, was crossed to a male carrying recessive genes for all the seven characters. Among the progeny, a female occurred (Fig. 53) that was hairy, scarlet and spineless on the left side of the thorax and head, and that was Dichaete, peach and sooty on the right side of the body. The division line passed through the right eye, which was scar-

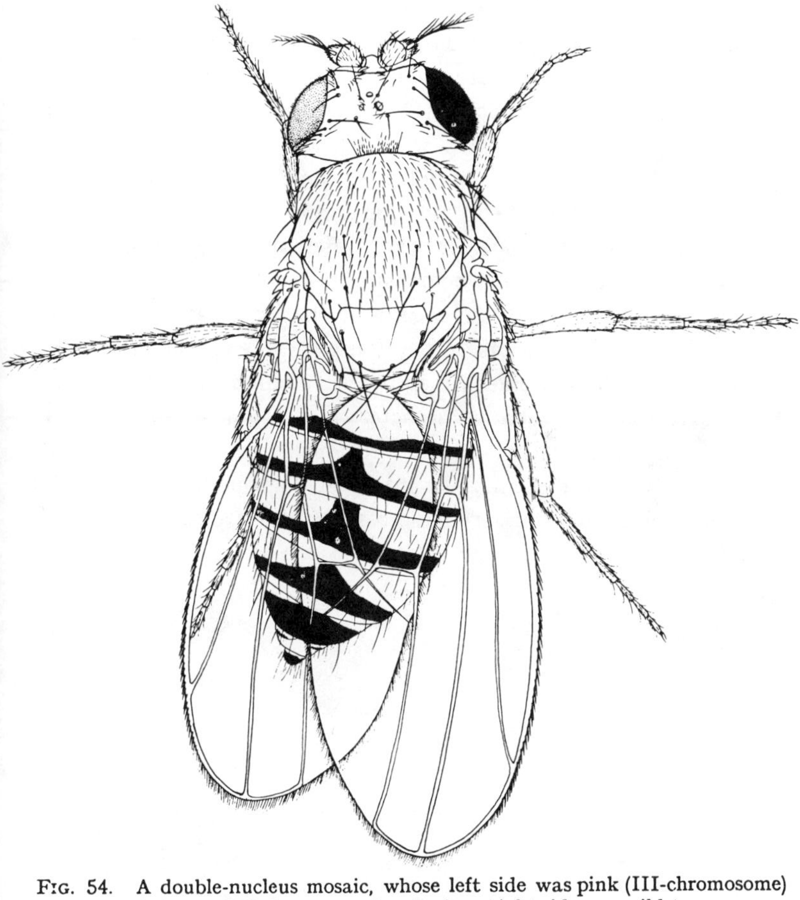

FIG. 54. A double-nucleus mosaic, whose left side was pink (III-chromosome) and shaven (IV-chromosome) and whose right side was wild-type.

let in the upper part and peach in the lower part. All of the abdomen belonged with the Dichaete, peach, sooty, side. The two sides differed in six characters. The fact that the peach region of the right eye was not roughoid is in conformity with the origination of the mosaic from a binucleated egg; for, the loss of the roughoid is due to crossing-over in the nucleus that gave the Dichaete, peach, sooty, pronucleus, while the hairy, scarlet, spineless, chromosome of the other pronucleus is a non-crossover. Each of these pronuclei was fertilized by a separate X-bearing sperm (both sides female) of the multiple recessive father.

A mosaic in which certainly two different sperms, and probably two female pronuclei, were involved, occurred in the F_1 of a cross between two second-chromosome recessives, vestigial and speck. The mosaic was vestigial and male on the left side and speck and female on the right side. The left side, accordingly, came from a pronucleus carrying the second-chromosome with speck and from an X-sperm carrying the speck second-chromosome. The right side came from a pronucleus carrying the vestigial second-chromosome and from a X-sperm with the vestigial second-chromosome. Thus, two different pairs of chromosomes, the X and the second, were demonstrably involved. In another mosaic one side was pink and shaven and the other side not-pink and not-shaven (Fig. 54). In this case, then, there is evidence that the third- and the fourth-chromosomes were involved simultaneously. It is only rarely that a cross involving characters in different groups is made, and the definite demonstration that all four chromosomes are simultaneously involved in the mosaics interpreted as due to two pronuclei and two sperms must come from a cross specifically designed to check all four chromosomes.

CHAPTER XV

TRIPLOIDY, INTERSEXES, SUPERSEXES

The intersexes

In an experiment made by BRIDGES to determine more closely the locus of the second-chromosome recessive "brown" by means of a three-point back-cross involving, besides brown, the well-known second-chromosome recessives plexus and speck, one culture was found that produced a total of 96 females, 9 males and about 80 individuals that were intermediates between females and males. (BRIDGES '21*c*).

The "intersexes," which are easily distinguished from males and from females, are large-bodied, coarse-bristled flies with large, roughish eyes and scalloped wing margins. They show marked fluctuation in all characters; and, in addition, apparently form a bimodal group with respect to the more strictly sexual characters — on the one hand a more "female-type", and, on the other, a more "male-type".

In the "female-type" (Fig. 55) the genitalia are predominantly or entirely female, and the abdomen is more nearly female in shape and color and in most of its morphological features. Spermathecae are present. The gonads are typically rudimentary ovaries; but the tissue in the terminal chambers of the ovaries resembles spermatogonial tissue more closely than oogonial tissue. Occasionally a gonad is mainly ovarian in appearance, but with a testicular bud. Sometimes also one gonad is a rudimentary testis and the other a rudimentary ovary. Sex-combs, a male character, are almost invariably present on the tarsi of the forelegs, although they are sometimes absent from one tarsus or reduced in size.

In the "male-type" (Fig. 56) the color, shape, and morphology of the abdomen are predominantly or entirely male, while the genitalia are mainly of the male type. Sex-combs are always present. The gonads are typically rudimentary testes; but in the more extreme individuals, the testes are nearly normal in size and contain bundles of sperm. Both types of intersexes have been found entirely sterile.

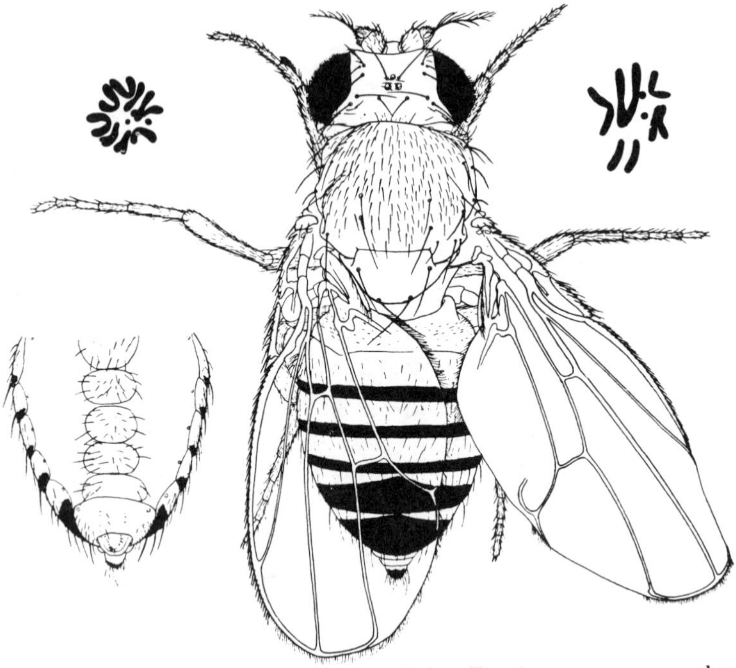

FIG. 55. Female-type intersex, with ventral view. The chromosome groups show two X's, three each of the large and of the small autosomal V's, and usually three small round IVths. In addition the group to the left had a Y-chromosome. The difference in size between the autosomal V's is especially great in these figures.

The culture that produced the intersexes showed other striking peculiarities; for example, the females and males showed three, instead of two, large classes representing original combinations. These classes were plexus speck, plexus brown, and brown speck. Extensive breeding tests were made of these flies, and each was found to have received from the father, a second-chromosome carrying the three recessives, plexus, brown and speck, which agreed with expectation. But from the mother each had received one of three different chromosomes; namely, a plexus speck, a plexus brown, or a brown speck chromosome. That is, the mother of the intersexes had carried three second-chromosomes, instead of two. For each of its loci, plexus, brown and speck, she had carried two recessive genes for the mutant character and one wild-type allelomorph, with nearly complete dominance of the wild-type gene in each set.

A condition of triploidy for certain sections of chromosome had been met with in the previous (unpublished) studies on duplications and on

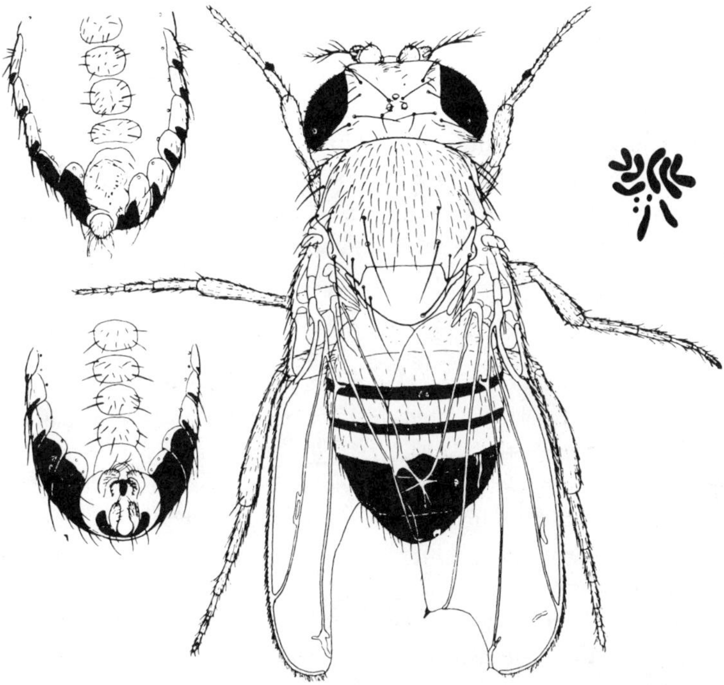

FIG. 56. Male-type intersex with ventral view to the left, and a ventral view of an intermediate specimen above. The chromosome groups show two X's, three seconds, three thirds, and usually only two fourths, and not three as shown here.

translocations; but that this triploidy was far more extensive soon became evident. The third-chromosome recessive, white-ocelli, had been present in the original culture; and tests of the flies produced in that culture showed that white-ocelli had been distributed in the same abnormal ratio as plexus, brown and speck. Not only were the second- and third-chromosomes involved, but the X as well, as was shown by specific tests with sex-linked characters. That the intersexes were themselves triploid and not diploid with respect to the autosomal characters was evidenced by the fact that the classes with respect to plexus, brown, and speck in the intersexes were strikingly different from those in the males and females; the intersexes showed three large classes also, but these were the single recessives plexus, brown, and speck, respectively.

The hypothesis that triploidy was involved was easily put to test by direct cytological examination. The chromosomes of the intersexes consisted of two sets of three V's (the two sets differing in the size of the V's), of a pair of rods, of three or two small round chromosomes, and usually of a J-shaped chromosome. That is, all intersexes possessed the second- and the third-chromosomes in triplicate and the X in duplicate, but they might lack one of the expected three fourth-chromosomes and have or lack a Y-chromosome. Thus four sub-types of intersex constitution were found (Figs. 55 and 56).

The 3N females

About ten percent of the daughters from the original intersex-producing culture, when tested, in turn produced intersexes and further disturbances of the ratios. These females were, by inference, triploid (Fig. 11*b*) for all the chromosomes (except the fourth, which might be present in duplicate). It was then discovered that these intersex-producing females could be identified by their somatic characters which together constitute a distinct type. They are most readily identified by large, coarse-textured eyes, and, secondly, by large body-size and stocky build. The sexual characters are completely female. They are very fertile. The productivity is low, because, as will appear, a very high proportion of the zygotes are inviable combinations. Genetic tests of these intersex-producers show that they possess in triplicate the loci for a large variety of first, second, and third group mutants. This genetic proof of triploidy was checked by direct cytological examination, and it was found that the intersex-producers were 3N in constitution. This finding extends the direct proof of the chromosome theory of heredity to specific second and third group mutants and corresponding specific V-shaped chromosomes. Such direct proof had already been provided by the data on non-disjunction for sex-linked mutants and the rod-like chromosome, and, more recently, for the fourth-group mutants and the small round chromosome.

The intersexes and 3N types lead to the conclusion that sex in

Drosophila melanogaster is determined by the autosomes as well as by the X-chromosomes; the ratio of autosomes to X's being the significant relation. The old formulation of 2 X = ♀ is at once seen to be inadequate; for, here we have individuals that have two X-chromosomes and yet are not females. They are shifted out of the female class by the presence of an extra set of autosomes, andthere by the autosomes are proved to play a positive role in the production of sex. Since the intersexes dfifer from females by the assumption of certain male characters, this effect of the autosomes is due to an internal preponderance of "male-tendency" genes. We may now reformulate the sex-relations as follows: both sexes are due to the simultaneous action of two opposed sets of genes, one set tending to produce the characters called female and the other to produce the characters called male. These two sets of genes are not equally effective; for, in the complement as a whole, the female-tendency genes outweigh the male, and the diploid (or triploid) form is female. The male-tendency genes in the autosomes are more numerous or more effective than those in the X-chromosome, while the net effect of genes in the X-chromosome is a tendency to the production of female characters. When, in a diploid zygote, the relative effectiveness of the female-tendency genes is lowered by the absence of one X, the male-tendency genes outweigh the female, and the result is the normal male. When the two sets of genes are acting in a ratio between these two extremes, as is the case in the ratio of 2X : 3 *sets of autosomes*, the result is a sex-intermediate — the intersex.

The Supersexes

The phenomenon of intersexuality might be expected to have a reciprocal phase, namely, supersexes. If the intersexes result from an intermediate ratio of X to autosomes, then when the normal female ratio of X to autosomes is exceeded, a superfemale might be expected to result; and, by increasing the relative number of autosomes, a corresponding supermale might be produced. Diploid individuals possessing an extra X have been identified as a distinct somatic type that is ordinarily lethal. (Fig. 57). This type resembles the female, but has abnormal ovaries and is sterile. Conversely, individuals with one X-chromosome and an extra set of autosomes have been identified. (Fig. 58). These resemble the males, but form a distinct type, and are sterile.

As stated, the intersexes generally show two types, one more female-like, the other more male-like. At first it was surmised that

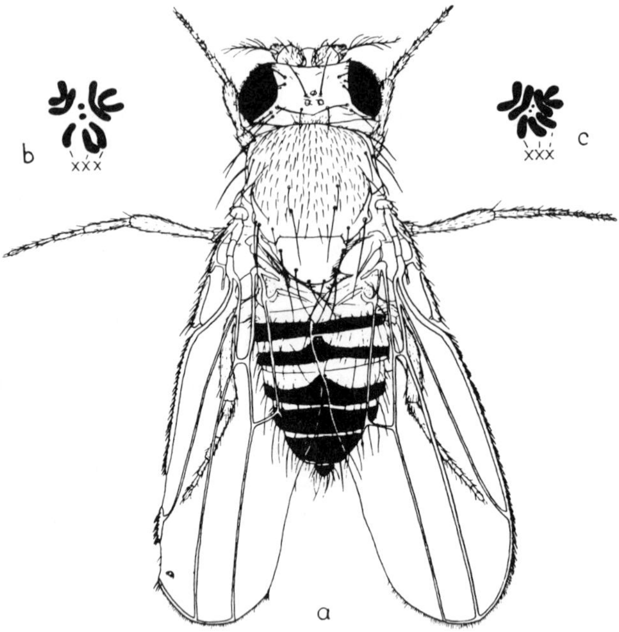

FIG. 57. A triplo-X, or Superfemale, with two chromosome groups showing the extra X-chromosome.

the female-type corresponded to the group with two instead of three fourth chromosomes. More recent results have shown that usually the female-type has three IV-chromosomes and the male type two (Table 7).

TABLE 7. — SERIES OF SEX-TYPES IN D. MELANOGASTER

Sex		X's	Sets autosomes	Sex-index
Superfemale		3	2	1.5
Female	4N ♀	4	4	1.
	3N ♀	3	3	1.
	2N ♀	2	2	1.
Intersex	♀-type	2	3	.67
	♂-type	2	3(−IV)	.67—
Male		1	2	.5
Supermale		1	3	.33

As SCHRADER and STURTEVANT pointed out ('23), by assigning negative values to X and positive values to "set of autosomes", and by assuming that the interactions of these two sets is by summation, a

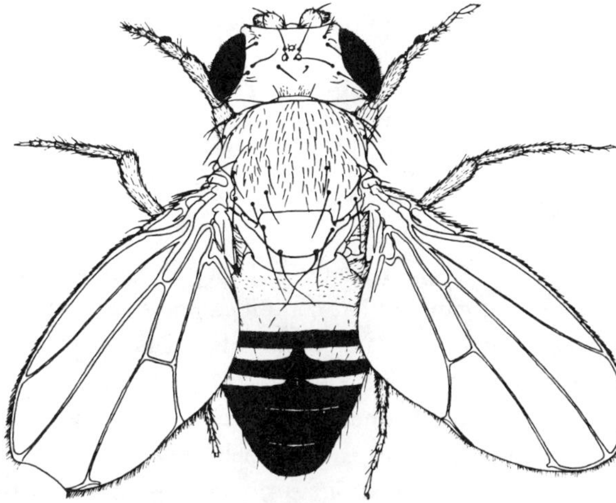

FIG. 58. A Supermale. Recent cytological examination shows one X and three of each autosomes.

series of sex-indices can be derived that will allow a haploid individual to be male, while on the system shown in the above table a haploid (1N) individual would be expected to be female.

The Gametic Series in 3N Females

Since each kind of chromosome exists in triplicate in the 3N female, the gametes with respect to a given chromosome (C) are expected to be 2C and 1C with equal frequency. Thus, half the gametes should be haplo-X and half diplo-X. If the reductions in the different kinds of chromosomes are independent of one another, then, among the haplo-X gametes, half are haplo-II and half are diplo-II, and so on for chromosomes III and IV. Disregarding the fourth chromosome, there should be eight types of gametes, including the extremes, N and 2N. These are fertilized by two types of sperm, the X- and the Y-sperm. Thus, an N-egg by an N-sperm (X-sperm) results in a 2N-female. But an (N + X)-egg by a Y-sperm also results in a female (XXY). These two types of females can be distinguished by mating the 3N female to a male with the sex-linked dominant Bar, for the N-eggs then give Bar daughters while the (N + X)-eggs give wild-type daughters.

By such a mating the frequency of the various types of gametes can be investigated. Table 8 gives the results of a recent cross of this type carried out by BRIDGES (unpublished).

TABLE 8. — PROGENY FROM 3N ♀ × BAR ♂

Zygotes	♀	B ♀	♂	B 3N♀	Ix	B Ix	B S-♀	S-♂
Totals	1,116	154	162	105	104	898	36	142

These data agree with other results in that no zygote can be identified as coming from a gamete that contained one second chromosome and two thirds, or two seconds and one third. The eggs that do give rise to viable zygotes may be divided into two classes, based on the type of reduction division. In one type 2 chromosomes of each sort pass to one pole and one of each sort to the other. From this type of reduction there result two kinds of mature eggs, equally numerous, (N and 2N). In the other type of maturation, two sets of autosomes and one X pass to one pole, one set of autosomes and two X's to the other. There result two equally frequent types of eggs (N + A and N + X). In some of the classes of zygotes very high mortality is involved, as in the supermales and superfemales. In the intersexes and 3N type the mortality is considerable. The only two classes free from abnormal mortality are the wild-type female and the wild-type male classes, which were 1116 and 162 respectively. Normal males have a viability of 95 as compared with 100 for the normal females, so 162 should be corrected to 171 if it is to be compared to the number of females, 1116. The ratio of 171 : 1116 is 1 : 6.5, which is the ratio of N : N + X gametes. This ratio agrees, when allowance is made for the known mortalities, with the ratios shown by all other classes coming from the N and 2N gametes on the one hand and the N + X and N + A gametes on the other. On chance distribution of chromosomes the members of these two categories should be to each other in the ratio 1 : 1; that is, in the 3N ♀ the chromosomes are *not* distributed according to chance, but the distribution of autosomes affects the distribution of sex-chromosomes or *vice versa*.

Crossing-over in the 3N female

In cooperation with E. G. ANDERSON, BRIDGES has made a study of the crossing-over in the X-chromosome of the triploid female (unpublished data). Each of the three X-chromosomes is followed by means of five mutant loci. For example, one of the five points was marked by yellow in one X, by its wild-type allelomorph in a second X, and by scute, which gives no crossing-over with yellow, in the third. In addition, each X carried a recessive mutant gene, crossveinless. Such a 3N female was outcrossed to a not-crossveinless male. All the crossveinless daughters of this outcross have received from the mother two X-chromosomes that have emerged from the same reduction division. A study of the characters representing the five marked points present in such females shows the origin of the different parts of each of the two X-chromosomes. It has been found that crossing-over may occur between all three X-chromosomes simultaneously, since chromosomes emerge that are composed of parts taken from each of the three chromosomes that entered. The most frequent type of female is one that carries one chromosome that has not undergone crossing-over (and is a duplicate of one of the entering chromosomes) and another X that is composed of parts of the other two. A fairly frequent result is the emergence of two X's that are duplicates of each other for part of their length but not for the remaining part. This result indicates that each of the original chromosomes was in a two-strand stage when crossing-over occurred. In these 3N females crossing-over is relatively less frequent in the right end (from lozenge to Bar) than it is in the normal (2N) female, while in the left it is relatively much more frequent. Thus, about 15 percent of the emerging chromosomes have come from crossing-over between yellow and ruby, while the normal percentage is about 8. In the extreme left end crossing-over has been observed between yellow and scute, which had not been certainly detected in the diploid form. A more extended analysis of the data is expected to throw light on the details of the process of crossing-over.

Frequency and mode of origin of triploidy and of tetraploidy

Since the discovery in 1920 of the intersexes on which these studies were based (BRIDGES had found other intersexes earlier but had not been able to continue the strains) intersexes of the triploid type have been found 15 times and there have been 7 discoveries of 3N females originating *de novo* from 2N lines. This very high frequency for the occurrence of triploidy is paralleled by cytological observations that give the explanation of the origin of triploidy. In three different preparations of 2N females BRIDGES found that portions of an ovary were constituted of markedly larger cells, and in two of these cases the chromosomes could be counted as 4N. Evidently there had been division of the chromosomes in some oogonial cell, followed by non-division of the nucleus. The resulting tissue was tetraploid, and all reduced eggs (or sperm, if the non-division occurred in the male) would be diploid. These gametes, uniting with normal haploid gametes from the other parent would give rise to 3N zygotes. When a 2N egg is met by a Y-sperm the result is an intersex; likewise when an XY2A-sperm meets an N-egg the result is an intersex.

In two preparations of intersexes BRIDGES found similar cysts, this time containing even larger cells, in one of which the chromosomes are clearly 6N (except for the X's). The intersexes are sterile; but such non-division occurring in a 3N female would give rise to 6N cysts and 3N-eggs, which would give 4N zygotes. In the breeding work with 3N females there is a chance to secure such 4N zygotes. As a matter of fact, BRIDGES found two cultures, and L. V. MORGAN one, in which supposedly 3N females descended from 3N mothers, gave only triploid offspring, viz., 3N females and intersexes, in equal numbers. The eggs of the 4N females were all 2N, which gave 3N females with X-sperm and intersexes with Y-sperm. Tetraploid females, on the basis of these three cases, resemble externally 3N females. They are expected to be even larger, and with the characteristics of the 3N individuals still further emphasized. The fact that 4N individuals are females that are not modified, adds evidence that the interaction of the sex genes in D. melanogaster, conforms to the "ratio type" (as shown in table 7), rather than to the "arithmetical type" (as proposed by Schrader and Sturtevant). Thus, the expectation is strengthened that a haploid individual of D. melanogaster would be a female.

CHAPTER XVI

Deficiency

The term "deficiency" has been used (Bridges '17) to designate "the loss or inactivation of an entire, definite, and measurable section of genes and framework of a chromosome." All the phenomena met with in the various cases in which the deficiency involved a section of a chromosome have been met with also in cases in which the deficiency involved a whole chromosome, e. g., in Diminished. In the case of Diminished the phenomena are clearly due to loss and not to inactivation. It has seemed best to subdivide these cases of deficiency according to the unit involved. In one class the unit is a section of chromosome and in the other class a chromosome (Bridges '22). The classes will be called respectively sectional-deficiencies and chromosome-deficiencies.

The reverted-Bar (page 29) may be regarded as a deficiency in which the unit is a single gene, and be included in this classification as a case of gene-deficiency. The term deficiency is applied to an end result rather than to whatever process produces this result, which is, in the main, non-disjunction in the case of chromosome-deficiencies, translocation in sectional-deficiencies, and unequal crossing-over in gene deficiencies. The term deficiency, used without qualification, may refer to the whole field, or may be used in its original meaning, which is equivalent to the longer term sectional-deficiency.

The deficiency for one and for both fourth-chromosomes has already been treated (page 138), and the case of the male, which may be regarded as deficient for one X, is discussed in several preceding sections. The deficiencies have a reciprocal aspect in the duplications (sectional-duplication, chromosome-duplication and gene-duplication) which will be discussed in Chapter XVII. The first sectional-deficiency found, involved the section between, and including, the loci for forked and Bar in the X-chromosome (BRIDGES '17), and is therefore called (f-B)-deficiency.

Forked-Bar Deficiency

A male that carried the sex-linked recessive white and the sex-linked dominant Bar was mated to a female that was heterozygous for eosin, which is an allelomorph of white. Besides the expected Bar daughters, there was present one exceptional daughter that was not Bar. But her eye-color showed that she had inherited the white from her father in spite of not having inherited his Bar. When she was outcrossed to a wild male none of her offspring were Bar, again showing that the Bar had not been inherited. The sons were of two classes, eosin and white, which again showed that white had been inherited. But these sons were only half as numerous as the daughters; and the missing sons, as shown by the linkage relations with white, were those that had received the white not-Bar X from the father. This lethal action showed that not simply the Bar gene had disappeared, but also one or more genes vital to the life of the fly.

If the non-inheritance of the Bar gene and of the vital allelomorphs was due to the loss of a section of chromosome including their loci, then all the other genes within a definite distance of Bar should also have been lost, and the female carrying one deficient X should behave as though haploid and not diploid for all mutants whose loci are included. That is, any recessive character whose locus is opposite the deficient-region should show itself in the female that has one deficient X and one X carrying the recessive gene. By this test for "pseudo-dominance" it was found that the deficient section was long enough to include the locus of forked, mapped as half a unit to the left of Bar, but did not reach to rudimentary, 1.9 units to the left of Bar, nor to fused, 2.5 units to the right of Bar.

It was found that in females carrying (f-B)-deficiency there was no crossing-over between the loci for forked and Bar. This means that not only the genes were effected but also the connecting structure or "framework" of the chromosome.

In linkage tests with females having one deficient X it was found that rudimentary and fused showed a distance of 3.7 units instead of the 4.4 normally found. That is, the map is shortened by an amount corresponding to the length of the deficient section; but crossing-over in the region immediately adjacent is apparently not affected.

One conclusion from the (f-B)-deficiency case has marked interest in connection with a similar conclusion from Bar-reversion (see page 32). It was concluded that the broadness of the eye of the normal heterozygous Bar female is not due to the action of the wild-type allelomorph, as formerly supposed, but to the fact that one narrowing gene (B) is opposed by two sets of broadening genes, one situated in each of the X-chromosomes in some other region than that involved in (f -B)-deficiency. Thus, it was found that the female carrying Bar in one X and Bar-deficiency in the other has an eye intermediate between the narrow eye of homozygous Bar and the round eye of the wild-type. This eye was indistinguishable from that of the normal heterozygous Bar; and since its broadness must be ascribed to broadening genes outside the region that has lost the Bar locus, and since these sets of broadening genes are likewise present in the normal heterozygote, it was concluded that the broadness in that case also must be attributed to them rather than to the action of the "round" allelomorph of Bar. The homozygous Bar female and the Bar male have a similar narrow eye from possessing these Bar-broadeners and Bar-narrowers in the same ratio (2B : 2 sets broadeners = hom. B ♀; 1B : 1 set broadeners = B ♂). That the Bar eye of the male is not quite as narrow as the Bar eye of the homozygous female may be due to there being a slight preponde rance in effectiveness of Bar-broadeners over Bar-narrowers in the autosomes.

An XX-individual having one (f-B)-deficient X is a normal female, which is in harmony with the view that sex-differentiation is dependent upon specific genes, situated in some part or parts of the X other than the section from forked to Bar.

That the maps for the distribution of genes along the chromosomes, derived from linkage studies, correspond to real distributions, at least as far as serial order is concerned, received strong confirmation from the (f-B)-deficiency. For, the various characteristics encountered in the case just discussed all find an adequate explanation in the hypothesis that a definitely *localized* section of the X-chromosome and its contained genes have been simultaneously affected by a single agency. The alternative hypothesis that these particular genes (Bar, the vital allelomorphs, forked) were affected, because of their possession of some common property other than location, and unshared by other genes, has no apparent support in view of the very diverse nature of the dominant Bar eye-shape, the recessive forked bristle-modification, and the lethals, and is directly negatived by the phenomena connected with the IV-chromosome deficiency, and by the deficiency originating from Pale-translocation (see page 176). The deficiency mutation was first discovered because of its effect upon a single gene, Bar. It was then found to have affected at the same time one or more vital allelomorphs. When these vital allelomorphs were mapped, according to the linkage shown, they were seen to occupy the same region as Bar. When forked was found to be involved the argument for a common locality became still stronger, for this relation was detected and proved as the result of a deliberate search among those genes that had *previously been mapped closest to Bar.* But the deficiency mutation affected not simply a section of genes mapped as adjacent, it at the same time removed crossing-over from a definite section of the chromosome. Now, when the section from which crossing-over had been removed is compared upon the map with the section in which the genes were affected the two are seen to be identical. All these coincidences, which occur likewise in other cases of deficiency, are unaccountable except on the ground that the maps correspond section by section with the real localization of genes.

Vermilion-deficiency

The first indication of "vermilion-deficiency" (BRIDGES '19) was a 2 : 1 sex-ratio, that showed that a new sex-linked lethal had arisen from ancestry free of lethals. The linkage relations showed that the locus of the lethal was in the neighborhood of the sex-linked recessive, vermilion, whose locus is at 33.0. When females heterozygous for the lethal were outcrossed to vermilion males, all the daughters that received the lethal-bearing chromosome showed vermilion eye-color, although, from the pedigree, vermilion was known to be absent from the ancestry. The lethal action and the "pseudo-dominance" of vermilion together showed that a deficiency had occurred for a section including the locus of vermilion and one or more genes necessary for the life of the individual. Linkage tests showed that the amount of crossing-over in the neighborhood of the deficiency was probably cut down, a feature not found in (f-B)-deficiency. Part of the five units difference found, was presumably due to the elimination of crossing-over throughout the deficient region, which is of unknown but small extent.

The Notch-deficiencies

The mutant character "Notch" was first found by DEXTER ('14), who showed that Notch was sex-linked, dominant in the female, and lethal in the male. In the next year BRIDGES found a new Notch (Notch-1) of independent origin, but agreeing in all known characteristics with the Notch of DEXTER. BRIDGES found that the locus of this Notch, and also the locus of another Notch (Notch-3) found by MORGAN, was about 1.5 units to the right of that of white in the X-chromosome.

Several other Notches were found, which agreed with those previously studied. One (Notch-6) showed a new characteristic of Notch, viz., pseudo-dominance for facet (METZ and BRIDGES, '17). The loci of the Notches had been found to be close to the right of white and therefore presumably close to the locus of the sex-linked recessive facet. When BRIDGES crossed Notch-6 to facet it was found that all the Notch daughters were at the same time facet, although no facet had been previously present in the ancestry. This pseudo-dominance at once suggested that the Notch mutation was in reality a deficiency.

The previous cases of (f-B)-deficiency and vermilion-deficiency had not shown any character changes in the female heterozygous for the deficiency. But if the Notches are deficiences, then a rather large complex of characters must be attributed to the loss of this particular section of X-chromosome. As already explained in detail in connection with the character changes of Diminished, which is due to deficiency for the fourth-chromosome (page 137), a deficiency is expected, on the theory of genic balance, to produce a character change whenever the ratio of plus to minus modifiers within the section is different, for any character, from that of the whole normal group.

The degree of development of Notch-6 and also of facet was markedly greater in these Notch-facet daughters than was the case in the original Notch female or in the facet female. This new phenomenon has been termed "exaggeration" and it also finds its explanation in the nature of the genes included in the region removed by the deficiency.

These genes had in their net effect a greater tendency to make the facet eye like the wild-type than to make it more extreme, and hence their loss allowed the remaining genes, which have a net effect in the opposite direction and of like degree, to shift the development in the direction of a more extreme facet eye. The F_1 Notch-6-facet females were also exaggerated with respect to infertility to such an extent that not one of the eighteen females tested produced offspring, and the line terminated abruptly.

One other Notch (Notch-4, found by MORGAN) was still on hand and BRIDGES crossed it to facet, and found the same pseudo-dominance of facet and the same exaggeration of facet and of Notch.

Notch-8, found by MOHR in 1918, has been the object of a very extensive study (MOHR '19, MOHR '23). Notch-8 showed all the phenomena encountered in the previous Notches, but differed from them in that it involved a longer section. Thus, Notch-8 gave pseudo-dominance with white and its nine allelomorphs, located 1.2 units to the left of facet, and with Abnormal 1.8 units to the right of facet — an interval of approximately 3 units. That the section is not longer than 4.7 units is shown by its failure to give pseudo-dominance with prune, which is just to the left of white, and with echinus, which is just to the right of Abnormal. Very extensive linkage tests showed that crossing-over is entirely eliminated from the deficient section, and that the map of the deficient X is 3.8 units shorter in that region than the standard map. The crossing-over in other parts of the Notch-X is standard in amount.

The included mutants — white and the white allelomorphs, facet and Abnormal — are all exaggerated. The grade of exaggeration in the eye-colors is greater than the dilution due to heterozygous white, i.e., the deficiency corresponds to an ultra-white.

Notch-8 also exaggerates certain characters outside of the deficient region — notably all light eye-colors.

A cytological examination did not reveal any certain abnormality in the X-chromosomes. A suspected narrowing of the inner end of the X is discounted by later evidence (page 134) which shows that the Notch end of the X is the outer free end.

About twenty-five occurrences of Notch have been recorded. All those tested involved a deficiency for facet, but only four involved a piece extensive enough to include white. One of these latter (N-18) gave a map shorter in the affected region than the N-8 map, but there is an indication that in the N-18 case, adjacent regions are reduced somewhat in the amount of crossing-over. This conclusion was drawn from a comparison made by Mr. LI upon the amount of crossing-over shown by four distinct Notches, two of which involved white (N-8 and N-18), and two did not (N-11 and N-19). LI showed also that the death of Notch males occurs in the egg stage.

Recessive lethals as possible deficiencies

It was shown that (f-B)-deficiency and v-deficiency had a recessive lethal effect. This raised the question as to what proportion of the other recessive lethals, which were numerous, might be due to unrecognized deficiencies. These might be detected by the test for pseudo-dominance with mutants mapped as adjacent, or possibly by an effect upon linkage relations.

The sex-linked recessive lethal No. 2 (l_2) was crossed by BRIDGES to the recessive mutant club, whose locus was very near to the locus of the lethal. The F_1 daughters that received the lethal-2 chromosome all were club (METZ and BRIDGES '19). Besides this pseudo-dominance for club, linkage studies showed that the map corresponding to the female heterozygous for lethal-2 is some five units shorter in that region than the standard map (MORGAN and BRIDGES '16). There is one fact somewhat opposed to the interpretation of lethal-2 as a deficiency, viz., the fact that very occasionally a lethal-2 male survives, and is an extreme club type that is sterile. This favors the interpretation of lethal-2 as an extreme mutant allelomorph of club, but it leaves unexplained the associated change in crossing-over. There seems no good reason to suppose that very short deficiencies, or deficiencies in particular sections, not including essential genes, or in chromosomes highly balanced internally, as the X is known to be, must invariably be lethals.

Lethal-IIa, a recessive lethal located near the right end of the second chromosome (MORGAN and BRIDGES '19) has been shown by BRIDGES to be accompanied by a decrease in crossing-over throughout the second-chromosome (data given by MORGAN '21).

Dachs-lethal (MORGAN and BRIDGES '19) was first detected because of the apparent non-inheritance of dachs when both parents were supposedly heterozygous. It was then found that a recessive lethal was present that apparently occupied the dachs locus, or at least was so closely linked as to give no crossovers in the 5,000 flies examined. Flies carrying this lethal, when crossed to dachs, gave half of the flies dachs. Linkage tests showed that the dachs-black recombination per cent was changed from about 16 to about 11. The above facts agree best with the assumption that a deficiency for a section including the dachs locus had occurred, but they are not decisive against the alternative hypothesis of a very closely linked lethal and an independent crossing-over modifier.

Dominants that are recessive lethals as possible deficiencies

The Notch-deficiencies were dominant in heterozygous form and lethal in the male, which corresponds to a homozygous condition. This suggests that many of the other dominants that are lethal when homozygous (e.g. Truncate, Beaded, Dichaete, Hairless, (Fig. 59) Delta, and the Minutes) are also, in reality, deficiencies.

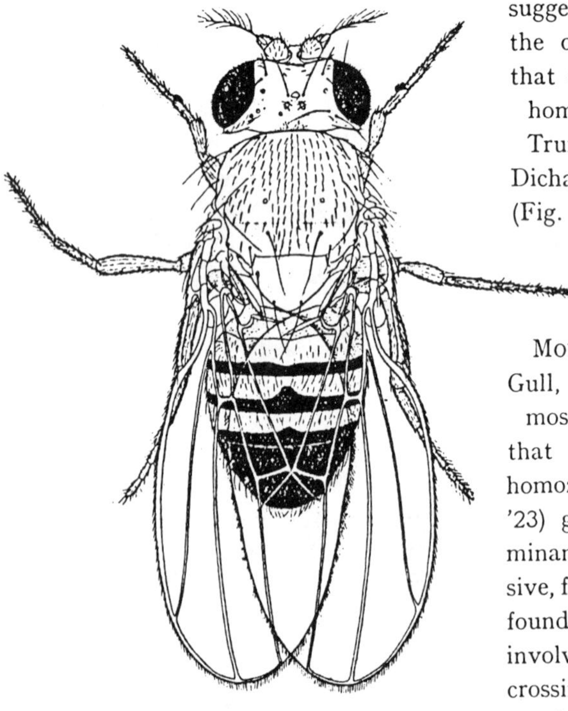

FIG. 59. Hairless. The place of the missing bristles is detectable by the basal rings that are not affected.

MOHR found that Gull, a second-chromosome dominant that is lethal when homozygous, (MOHR '23) gives pseudo-dominance for the recessive, fat; and BRIDGES found that in crosses involving Gull the crossing-over in that section was reduced.

Plexate, a dominant that is lethal when homozygous, was found to be situated in the right end of chromosome-II. When BRIDGES crossed Plexate to speck, all the F_1 Plexates were exaggerated Plexate and gave pseudo-dominance for speck. The linkage normality has not yet been tested.

Minute-1, a dominant that is lethal when homozygous, was found by BRIDGES to give markedly reduced crossing-over in the right end of the second-chromosome. Its locus was deduced to be in the neighborhood of plexus. When it was crossed to plexus it gave exaggerated M1-plexus compounds. Moreover, when it was crossed to arc, the locus of which is about a unit distant from that of plexus, it gave pseudo-dominance for it as well. This case, then, is very clearly a deficiency.

CHAPTER XVII

DUPLICATION AND TRANSLOCATION

The term "duplication" has been used by BRIDGES ('19) for cases in which a section of chromosome that is a duplicate of a known section is present in an abnormal location in addition to being present in its normal location. These cases are reciprocal to "deficiencies," in which a section is absent from its normal position without being present in some abnormal position. The term translocation (or transposition) was used (BRIDGES '19) for cases in which a section of chromosome is removed from its normal location but is present in an abnormal location.

Roughly speaking, duplications are additions, deficiencies are losses, and translocations are deficiencies neutralized by corresponding additions. It is apparent that a translocation gives rise, upon outcrossing, both to a deficiency and to a duplication. In duplication the fragment may lie free as an extra smaller chromosome, or be attached to another chromosome, or even to the same chromosome, e.g. a piece of X to the end of an X. The presence of an extra section gives rise to certain phenomena called "duplication-phenomena" including changes in Mendelian ratios, in linkage relations, and in dominance, also the production of character changes and of lethal effects, etc. These same phenomena are met with in case of the extra fourth-chromosome and the extra X-chromosome. It seems advisable, therefore, to subdivide the general field of the duplications, according to the unit involved, into chromosome-duplications, and section-duplications.

Vermilion Sable Duplication

In a back-cross test involving the three sex-linked recessives cut, sable, and forked, BRIDGES found that sable was appearing in about

20 % of the flies, instead of in 50 %, as expected. The sable was not being eliminated by a lethal; for the sex-ratio was normal, and so also were the relations of cut and forked. However, it was noticed that many of the flies that were expected to be sable were slightly darker than the wild-type; and accordingly it was assumed that these were flies in which some new situation had arisen which suppressed the action of the sable gene.

A cut, "suppressed-sable" male was crossed to a female carrying the two sex-linked recessives eosin and vermilion. In F_2 full sable reappeared, showing that the faintly sable male had carried the sable gene. As before, about 20 % of the flies were full sable, and about 30 % were faintly sable. A striking fact was that practically all of the full sable flies were at the same time eosin. That is, when the left end of the chromosome carrying the sable gene was removed by crossing-over, the suppressor was removed also. On the basis of the linkage results, the locus of the suppressor was calculated as to the left of eosin by an amount that should place it at the very left-most end of the chromosome.

In the above F_2 result a new abnormal feature appeared, viz., that in many flies vermilion was also being suppressed. Those vermilions were suppressed that had lost eosin by crossing-over and had at the same time gained the sable-suppressor. That is, the suppressor for the body-color sable is at the same time a suppressor for a very different character, the eye-color vermilion.

The body-color of the faintly sable males was about that of females heterozygous for sable, and accordingly it was assumed that those flies failed to be sable because they possessed, in addition to the sable gene, the not-sable allelomorph. The assumption was made that the suppressor was in reality a piece of the X-chromosome, carrying the wild-type allelomorph of sable, and that, in a former cross, it had been removed from its normal location in the middle of the wild-type X carried by a female heterozygous for sable and had been permanently joined to the end of the homologous X that carried the gene for sable. The presence of this fragment in a male would render him diploid with respect to the loci included in the duplicating fragment.

The loci included were, so far as tested, vermilion and sable, and not eosin, cut, nor forked. Subsequent tests showed that the recessives miniature, dusky and furrowed, whose loci are between those of vermilion and sable, were not affected. This offered a difficulty, that was met by the assumption that two fragments were translocated, either successively or simultaneously. This does not now seem so improbable as it did at the time the situation was encountered, for just such an independent duplication for the sable locus was found shortly afterward in another experiment; and the frequency of sectional-duplications is now supposed to be fairly high.

The presence of the duplicating fragment offered a new kind of opportunity to study dominance relations, for females triploid or even tetraploid for the vermilion and the sable locus could be bred. Thus, a male carrying vermilion in the normal locus and not-vermilion in the duplicating section is not-vermilion in appearance, like the normal heterozygous female. When such a wild-type male is crossed to a vermilion female all of the offspring are vermilion. That is, in the daughters the two recessive vermilion genes dominate over the one wild-type gene. The eye-color in this female is not quite as bright a vermilion as in the case where the fragment is absent. In a female both of whose X's carry vermilion in the normal locus and not-vermilion in the duplicating fragments the eye-color is again wild-type. That is, two wild-type genes dominate two recessive vermilion genes. Such a wild-type female crossed to a vermilion male gives crisscross inheritance of the Abraxas type, viz., the recessive vermilion entering from the father appears in all of his daughters and in none of his sons! Another aspect of the above cross is that a recessive character vermilion, is behaving as a dominant. Not only can a character involved in a duplication act as a dominant instead of as a recessive, and give the Abraxas instead of the Drosophila type of crisscross inheritance, but the locus can appear to be at the end of the chromosome instead of in the normal place. For, in a female that is homozygous for vermilion in the normal locus and carries one not-vermilion duplication, and is also heterozygous for other sex-linked characters such as yellow, cut, and forked, the linkage relations of vermilion agree with a position at the left end of the chromosome and not at the normal locus of 33.0. The presence of the duplicating fragment apparently does not change the amount of crossing-over along the chromosome.

A cytological examination of females, one of whose X's was normal and the other carried (V,S)-duplication, did not give a decisive difference, though slight differences in length seemed to be detectable in some cells. The recent evidence from the double-X's shows that the left end of the map corresponds to the free end of the X, and not, as was formerly thought to be the case, to the spindle-fiber end.

Sable-duplication

In a linkage experiment in which the mother was from the cross of vermilion forked to sable garnet $\left(\frac{v \qquad f}{[s\ g]}\right)$ there occurred a male that showed only garnet. For a garnet male to lose sable without at the same time gaining vermilion would require double crossing-over within the very short distance from vermilion to garnet. It was believed that this distance was too short to make double crossing-over at all probable as the explanation of the disappearance of sable. In F_2 from crosses with the exceptional male, full sable reappeared, showing that the apparently not-sable fly had really carried the sable gene. But only about 20 % of the F_1 flies were full sable, and about 30 %, when closely examined, were found to be faintly sable. The faintly sable flies were nearly all garnet eyed, and since there is only a little crossing-over between sable and garnet (1.4 units of distance), these not-sable garnet flies could be relied upon to be also really sable, but with the sable suppressed. In the F_1 of crosses with eosin the linkage relations showed that the locus of the suppressor was to the left of eosin, that is, at the left end of the chromosome. The subsequent crosses with the sable-suppressor gave results identical with those encountered in (V,S)-duplication, except that vermilion was not affected. It was assumed, then, that this was an independent origination of a duplicating fragment involving the sable locus and joined to the left end of the X-chromosome.

If the sable-suppressors in the two cases are each the wild-type allelomorph, then when a fly is made up that has sable in both normal loci, and one X carries not-sable from (V,S)-duplication, and the other X carries not-sable from sable-duplication, such a female should be wild-type. If, on the other hand, the suppressors are independent genes, such a fly should be sable, since each suppressor is recessive in the female. When the experiment was performed, it was found that the flies were wild-type, showing that the same agent was the suppressor in both these cases, viz., the wild-type allelomorph of sable.

Pale-translocation

Certain phenomena observed by BRIDGES in studying the eye-color "Pale" were explained as due to the transference of a piece from the right end of chromosome-II to an attachment near the right end of chromosome-III.

The zygote in which one deficient II-chromosome is accompanied by one third-chromosome carrying the fragment of second-chromosome is not distinquishable in appearance from a wild-type zygote. But if such a zygote is, in addition, homozygous for eosin, then the eosin color is diluted, and the "Pale" color can be used as an index of this constitution. When such a fly is outcrossed, one quarter of the gametes contain both modified chromosomes, and continue the "Pale" line. Another quarter receives the unmodified II and the unmodified III, and gives a completely normal line. A third quarter receives the deficient II-chromosome, and all zygotes from them die, presumably from too great a change in characters. The fourth quarter receives the third-chromosome with a duplicating section of second-chromosome. These gametes give zygotes that are 2N but that have an extra piece of II-chromosome. Such zygotes differ from the wild-type in having a squatty build, in having blunt wings, with the eyes somewhat smaller, and in other slight particulars. Their viability is only about 80 % that of wild-type flies. If such a fly carries the gene for the recessive character speck in each of its normal second-chromosomes it is nevertheless not-speck in character. The wild-type genes in the duplicating fragment act as dominant suppressors not only for speck but also for plexus, brown, morula, balloon, purploid, blistered, and lethal-IIa, that is, for all mutants that lie to the right of arc in the second-chromosome. This is an accurate measure of the length of the duplicating fragment. A second measure arises from the fact that a second-chromosome that has lost its end can not undergo crossing-over in that region with a normal II-chromosome. Experiments show that from a point between arc and plexus to the extreme right end of the chromosome crossing-over is eliminated. Flies homozygous for the duplicating fragment die. But if they have two duplicating sections and one deficient section, the net effect is one duplicating section, and these flies are found to be viable in about 80 % of the cases and to resemble closely in their character-changes the flies that have simply one duplication.

Flies homozygous for the deficient second-chromosome and at the same time for the duplication-bearing third-chromosome are lethal. This fact, together with the dilution effect upon eosin, is interpreted as due to the duplicating section being not quite as great as the deficiency; a short piece of the section was lost in the process of transference.

The point of attachment of the duplicating section was found to be on the right limb of chromosome-III, between ebony and rough. That this locus is not at the end (that is, not to the right of rough and other loci) nor at the spindle fiber at the middle (near pink) gives a clue as to the possible mode of origin of this translocation. It may be assumed that in the process of synapsis the two homologous chromosomes, originally fairly separated, came together and began to synapse at their free ends first (bouquet stage). This synapsis progresses along both limbs to the middle of the chromosome. According to the observations of Gelei ('20) it sometimes happens that the end of a chromosome of one pair lies between the two ends of another pair. Usually these entanglements are overcome; but if the entangled chromosome should not extricate itself, then the continuance of the synapsis would result in two pachytene threads interlocked by half the thickness of each at a single point near one end of each. If the second and third pair of chromosomes became thus interlocked, the subsequent manoeuvers of the chromosomes may be supposed to have put enough strain on these strands so that the end of the second-chromosome, that was pinched between the third-chromosomes, was torn off and carried away by the third-chromosomes. Or perhaps there was a fusion between the inner second-and third-chromosome strands at this point of mutual interlocking; and this fusion made a break imperative, and also accounted for the fact that the broken end remained attached to the third-chromosome at the point of contact.

A cytological examination seems to show that the shorter pair of V's has one arm abnormally short (the right end of the second-chromosome). But that the third-chromosome is increased, has not been established.

Dominance in Triploidy

These cases of duplication gave the first opportunity in Drosophila to study the effects of two recessive mutant genes acting in company with one dominant wild-type allelomorph. In the case of vermilion and of sable the two recessives almost completely dominate the one wild-type allelomorph; in the case of plexus in Pale-translocation, the two plexus one not-plexus form is plexus-like, but the proportional effect of the wild-type gene is greater than in the other two cases. In the case of brown, speck, blistered, morula, balloon, and lethal-IIa, the wild-type allelomorph allows very little if any effect of the two recessives to be seen. The triplo-IVths and triplo-X's later gave opportunity to study a wider range of characters. In the triplo-X it was found that two vermilions do not dominate one not-vermilion. The same holds for sable. Later still, the intersexes and 3N females gave opportunity to study any gene in triploid condition. In the 3N female it is quite generally true that two recessive mutant genes are overruled by a single wild-type allelomorph. This holds for vermilion, for sable, and for plexus. Here is an apparent contradiction to the findings in the duplications. The first conclusion is that dominance is not a simple property of this or of that gene, but is dependent upon other conditions as well. There are two differences to which the effect may be attributed. The triplo-X differs from the duplication case by the presence of nearly a whole X-chromosome instead of a very short section of a chromosome. The multitude of genes in this extra X may well be the cause of the difference. In the case of the 3N there is still a wider range of other genes that may be affecting the result. Another difference, at first considered relatively unimportant, has been emphasized as the result of STURTEVANT's demonstration that it is a large factor in the interaction of the genes of the Bar locus, namely the "positional effect." STURTEVANT found that two Bar genes in the same chromosome produced a greater effect upon reduction in eye-size than two Bar genes in opposite chromosomes. In the cases of duplication and of triploidy such chances for genes close together to overrule a third, more distantly situated gene, are present.

A second case of translocation

Another case has been found by BRIDGES in which the right end of the second-chromosome is attached to the third-chromosome. In this case the corresponding deficient second-chromosome is not on hand. The piece involved in the new case is apparently not as extensive as that derived from Pale-translocation, since second-duplication is viable in homozygous state.

The suppressors of tan, of scute, and of Hairy-wing as possible cases of translocation

In working with the sex-linked recessive tan, BRIDGES found that certain flies that should have been homozygous tan were nevertheless wild-type. Analysis showed that they were homozygous tan, but that tan was suppressed by a recessive in the third-chromosome, located to the left of Dichaete.

In working with the closely linked recessives scute and apricot, MORGAN found that many apricots failed to show scute. Here also the suppressor was recessive and located in the third-chromosome.

The character Hairy-wing is a sex-linked dominant. BRIDGES found a recessive suppressor for Hairy-wing that is located at 55 in the third-chromosome. The females that are homozygous for this suppressor are sterile. The homozygous suppressor type also differs from the wild-type in having a squatty body-build and blunt wings. These, and a few other cases of suppressors, are interpreted as duplications.

CHAPTER XVIII

Cytological and genetic comparison of the species of Drosophila

The genus Drosophila includes over 200 described species from all parts of the world, and there can be little doubt that further collecting in tropical regions will greatly increase this number. The species

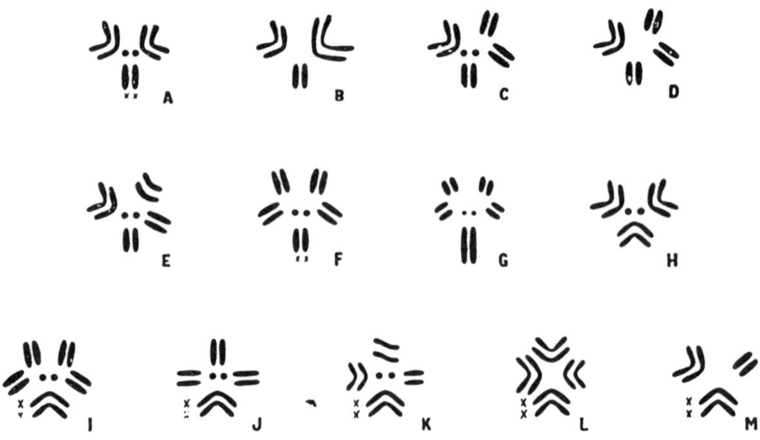

FIG. 60. Chromosome groups of species of Drosophila. (After Metz and Moses '23).

are apparently most numerous in the East Indies and in the Hawaiian Islands, but even in temperate North America, which has fewer forms of the group than has any other main geographical region, there are about 30 species. With no known exceptions, these species are perfectly definite and distinct from each other. Variable species do occur, but in no case that has been adequately studied is there any question as to where to draw the line between separate species. This is due to the facts that interspecific sterility has, so far, been found to be complete; and that a group of interfertile forms has not yet been found to show significant geographical variation, or correlated variation of separate characters.

The chromosome groups occurring in Drosophila and related genera have been studied by METZ ('14*b*, '16*a*), and by METZ and MOSES ('23). Fig. 60 shows the 13 different groups found by them. The various types are found in the following species: —

Type A

Drosophila busckii, COQUILLETT.
Drosophila bromeliae, STURTEVANT.
Drosophila florae, STURTEVANT.
*Drosophila melanogaster, MEIGEN.
Drosophila nebulosa, STURTEVANT.
Drosophila quinaria, LOEW.
Drosophila robusta, STURTEVANT.
Drosophila saltans ,STURTEVANT.
Drosophila simulans, STURTEVANT.
Chymomyza amoena, LOEW.
Chymomyza procnemis, WILLISTON.
Mycodrosophila dimidiata, LOEW.
Scaptomyza graminum, FALLÉN.

Type B

Drosophila earlei, STURTEVANT.

Type C

Drosophila calloptera, SCHINER.
Scaptomyza adusta, LOEW.

Type D

Drosophila immigrans, STURTEVANT.

Type E

Drosophila melanica, STURTEVANT.

Type F

Drosophila cardini, STURTEVANT.
Drosophila mulleri, STURTEVANT.
Drosophila ramsdeni, STURTEVANT.
DRosophila similis, WILLISTON
Drosophila tripunctata, LOEW
Drosophila virilis, STURTEVANT.
Drosophila species (Undescribed European species near D. obscura).

Type G

Drosophila funebris, FABRICIUS.

Type H

Cladochaeta nebulosa, COQUILLETT.

Type I

Drosophila hydei, STURTEVANT.
*Drosophila repleta, WOLLASTON.

Type J

*Drosophila obscura, FALLÉN.

Type K

*Drosophila affinis, STURTEVANT.

Type L

*Drosophila caribbea, STURTEVANT.

Type M

*Drosophila willistoni, STURTEVANT.

(*Only in the species marked with an asterisk has the X-chromosome been certainly identified).

As METZ has pointed out, it is possible that definite identification of the X-chromosome may cause some of these forms to be shifted to other types, as has in fact already happened for D. willistoni (see below). It is, however, hardly possible that this will happen in the case of any species that has yet been intensively studied genetically.

The type of chromosome group that appears to be almost universal in the other members of the enormous group of Muscidae (to which Drosophila belongs) is quite different from any of those here figured (STEVENS '08, METZ '14, '16*a*). It consists of 6 pairs of chromosomes, of which 5 are commonly more or less V-shaped, while the sixth and smallest is a short rod or is nearly spherical. In all cases where the X has been identified it is this smallest chromosome. The only form with this chromosome group in which mutation has been described is the cheese-skipper, Piophila casei LINNÉ. The meager data on this species will be summarized on page 203.

Mutations have been described in thirteen species of the genus Drosophila, though only five (melanogaster, obscura, simulans, virilis, and willistoni) have been intensively studied. Table 9 shows roughly the available information on these 13 species, as well as some of the more striking specific characters that they possess. Full descriptions of the "wild-type" characters are given by STURTEVANT (1921).

TABLE 9

Species	Type of chromosome group	Haploid number of chromosomes	Shape of X	Number of linkage groups known	Number of sex-linked genes	Number of autosomal genes	Length of Fly (in mm.)	General color	Number of acrostical rows	Number of egg filaments	Sex combs	Distribution
affinis	K	5	V	1	0	2	2.2	Blackish	6	2	1	Eastern United States
busckii	A	4	?	2	0	3	2	yellow	8	2	0	Cosmopolitan
caribbea	L	4	V	1	0	1	2	yellow	8	2	0	Tropical America
funebris	G	6	rod	3	3	4	2.5	brown	8	4	0	Cosmopolitan
hydei	I	6	?	2	2	2	2.5	gray	8	4	0	N. and S. America
immigrans	D	4	?	3	1	5	2.5	yellow	8	4	0	Cosmopolitan
melanogaster	A	4	rod	4	150	250	2	yellow	8	2	1	Cosmopolitan
obscura	J	5	V	5	28	12	2.2	black	8	2	2	Pacific States (Europe?)
repleta	I	6	V	2	2	1	2.5	gray	8	4	0	Cosmopolitan
similis	F	6	rod	1	0	1	2.2	yellow	6	4	0	Tropical America
simulans	A	4	rod (?)	3	15	25	2	yellow	8	2	1	N. and S. America
virilis	F	6	rod	6	18	23	2.5	black	6	4	0	United States
willistoni	M	3	V	3	30	9	2	yellow	6	2	0	Tropical America

The most obvious way of determining the nature of the differences between these various chromosome groups would be through the study of species hybrids. This study has been attempted many times, but so far the only hybrids obtained have been those between Drosophila melanogaster and D. simulans. These hybrids will be discussed below, in connection with the genetic studies on simulans; for the present we may note only that the hybrids are all completely sterile, and that the parent species have the same chromosome group (Type A).

Between 150 and 200 different kinds of cross-combinations between species have been tried. These include all possible combinations of the commoner species and of those extensively studied genetically. All combinations that seemed at all promising (because of similarity in appearance or of mating habits, or because cross-courtship was observed) have been tried very extensively. Except in the case of melanogaster and simulans, cross copulation has been seen in only two combinations (first observed by C. W. METZ.); D. obscura ♀ × D. melanogaster ♂ and D. affinis ♀ × D. melanogaster ♂. These have both been obtained many times, and that with obscura has been tried under many conditions (see STURTEVANT '21, page 9); but no offspring have been obtained. Dissections show that about half of the females that have cross-mated for 10 minutes or longer, contain sperm in their ventral receptacles, and in one case these foreign sperm remained active for at least four days after mating. These females lay eggs, but no sign of development is visible in them without sectioning; either fertilization does not occur, or development proceeds only a little way.

It has therefore been attempted to correlate the different chromosome groups by studying the mutant types that occur in the separate species. It was hoped that modifications of the same gene occurring in distinct species could be identified by their similarity in somatic effect, and that in this way a comparison could be made as to sequence of loci, correspondence between chromosome segments, crossover frequencies, etc. It must, however, be admitted that this method of attack has not so far been altogether successful—in fact, it almost seems as though it grows less and less promising as more and more data are collected. The nature of the difficulties encountered will be discussed below.

The studies of other species have, however, brought to light many interesting and important facts. It has become clear that melanogaster is not different from its relatives in mutability, and that in general the genetic behavior of all the species studied is very similar. Crossing-over has been found to be limited to females in all cases examined. Non-disjunction has been found to be of the same nature in simulans, virilis, and willistoni as in melanogaster. Perhaps the most important result is that the number of linkage groups never exceeds the haploid number of chromosomes, though in obscura (with 5) and virilis (with 6) it does exceed the number, four, found in melanogaster, thus furnishing another crucial test for the chromosome theory.

In the following discussion the species are arranged according to METZ's classification of chromosome groups that is given above. The available data on those species that have been little studied are given in somewhat more detail than in the case of the adequately known species. In certain cases (especially those of D. simulans and D. funebris) new data are here published for the first time, and these are presented in more detail than are the summaries of previously published results. For these reasons the space accorded to the various species is not proportional to the amount or importance of the available information about them.

CHAPTER XIX

Genetics of Drosophila simulans

This species is closely similar to Drosophila melanogaster in most respects, and, though of common occurrence in the eastern United States, was not recognized as distinct until 1919 (Sturtevant '19*b*, '20*d*, '21*a*). There are three definite diagnostic differences: the eggs of simulans have longer filaments; the eyes are larger and the cheeks narrower in simulans; and the male genitalia are different, though far more alike than those of almost any other two species in the genus. Simulans also averages slightly darker in color and stouter in shape than melanogaster, and newly emerged specimens have a more marked dorsal longitudinal sub-cuticular streak in the thorax. The two species differ in their geographical distribution. Melanogaster is to be found in all temperate and tropical parts of the world; simulans is strictly American, ranging from New Hampshire to Brazil, but apparently absent in the West Indies and on the Pacific coast of the United States. As to size, larvae, female genitalia, food-habits, mating habits, and length of life-cycle, no differences have been found between the species. Although no perfect division figures have yet been studied in simulans, enough has been seen of the chromosomes to leave little doubt that the female group, at least, is identical with that of melanogaster. This conclusion is borne out by the report of Bonnier ('24) that the single clear chromosome figure he was able to find in female hybrids between melanogaster and simulans was not different from a normal melanogaster figure.

Hybrids with melanogaster

It is possible to obtain hybrids between melanogaster and simulans (Sturtevant '20*d*) and only in this one case among the Drosophilidae, and apparently even in the entire order of Diptera. Even here, the cross matings are much less often successful than are matings within either species. All the data available indicate that the difficulty in obtaining hybrids is due largely to failure of cross-mating, and that this in turn is due to failure of the females, rather than the males, to react normally to the opposite sex of the foreign species.

It is very clear after a little experience that the cross, melanogaster ♀ × simulans ♂ is much more often successful than is the reciprocal, simulans ♀ × melanogaster ♂. Especially in the latter case, the frequency of success depends in part on the relative vigor of the two stocks used, so that no numerical data can be of general applicability. However, the difference between reciprocals is so great that there is never any overlapping. In the only two cases where any actual records are available, about 1 in 4 melanogaster females produced hybrid offspring by simulans males, while only about 1 in 50 simulans females produced offspring when mated to melanogaster males. Both of these tests were with stocks that were near the optimum for their respective combinations.

The hybrids produced may be of either sex, but males and females are seldom equal in number from a given mating. By the use of sex-linked characters in both species it has been found that types resulting from non-disjunction (see section on chromosome aberrations) have a different fate from that of the regular offspring. Table 10 shows the data obtained from all matings in which regular and exceptional (non-disjunctional) offspring could be distinguished:

TABLE 10

	Regular ♀	Exceptional ♀	Regular ♂	Exceptional ♂	Total
mel. ♀ × sim. ♂	3058	0	0	236	3294
sim. ♀ × mel. ♂	349	8	1206	0	1563

It is possible that the exceptional males from the second cross may survive, since the regular females survive only when the mother comes from certain stocks, and very few data are available from XXY females from such stocks.

That the results are not due to a changing of the apparent sex of any of the individuals is shown by the sex-linked characters; XX is always female, X is always male. In the case of melanogaster ♀ (without a Y) × simulans ♂, it has been found that the male larvae grow slowly, rarely pupate, and then form only abnormal puparia. In this case at least the missing class dies, and there can be little doubt that the other missing classes suffer a similar fate.

Bonnier ('24) has studied the histology of the hybrid larvae. His account is based on three pathological specimens, but he states that "a number of the hybrid larvae were perfectly normal" as far as he examined them. It seems quite possible that the abnormalities found in the three larvae described were accidental, rather than due to the fact that they were hybrids. At any rate, Bonnier failed to find any structural peculiarity of the hybrid larvae to which their death could be attributed.

Analysis of the data in Table 10 shows that no class survives unless it carries a simulans X chromosome. But if a simulans X is present, survival is usual unless a melanogaster X is also present and the combination is made in simulans cytoplasm (regular females from the second mating). Unpublished data show that in this special case the result depends largely on the race from which the simulans mother comes. Certain stocks consistently give no regular daughters when crossed to melanogaster males, while certain other stocks as consistently do give them.

All the hybrids are completely sterile. The development of the gonads is somewhat variable, but in no case are they normal in size. The females sometimes show no trace of oogonial cells in their minute ovaries, which then consist chiefly of tracheae and connective tissue. Bonnier ('24) has studied the ovaries of hybrid females in some detail. He reports them variable in degree of development, but in no case did he find eggs that were at all near the normal size of mature eggs.

Structurally the hybrids are intermediate between the parent species in all the differentiating characters (except egg filaments, which are not produced). They also often show certain defects (missing bristles, broken or missing crossveins, narrow wings, roughened eyes, or abnormal abdominal sclerites), which may be taken as indications of lack of coordination between the two gene complexes that are present. It may be noted that some of the same characteristics are found in certain of the types of melanogaster that have abnormal chromosome groups (see section on chromosome aberrations).

Linkage relations in simulans

Forty-two mutant types have been found and studied in simulans (STURTEVANT '21*a*, '21*d* and unpublished data). The genes concerned fall into three linkage groups: 18 are sex-linked, 12 lie in chromosome II, 12 in chromosome III. Chromosome IV is, cytologically, small like that of melanogaster; accordingly, it is not surprising that no mutant gene has been found to lie in it. Table 11 gives the names and loci of the mutant genes of simulans, together with the characters they affect.

TABLE 11

Chromosome I		
mutant	locus	character
yellow	0	body color
prune	3	eye color
dwarf	3	size
Gull	3	wings, hairs
ruby	12	eye color
crossveinless	15	wing veins
vesiculated	21	wings
singed	23	bristles
dusky	39	wings
garnet	44	eye color
tiny-bristle [2]	44.5	bristles
rudimentary	56	wings
forked	58	bristles
tiny-bristle	59	bristles
fused	63	venation
bobbed	68	abdomen, bristles
lethals	—	life

Chromosome II		
mutant	locus	character
black	0	body color
purplish	17	eye color
Shaven	17	bristles, abdomen
polychaete	30	eyes, bristles
spread	37	wings
plum	60	eye color
arc	59±	lethal
semi-forked	—	bristles
branched II	—	venation
kidney-II	—	eye shape
intersex	—	sex
frayed	—	wings, bristles

Chromosome III		
mutant	locus	character
curved	0	wings
dachs	20	wings, legs
scarlet	35	eye color
Deltoid	59	veins, hairs
bipectinate	73	legs, wings
peach	101	eye color
rugose	103	eye surface
Minute	129	bristles
roughish	—	eye surface
branched III	—	venation
kidney III	—	eye shape
lethal	—	life

Corresponding genes in melanogaster and simulans

Sixty-five recessive mutant genes of melanogaster, in 54 loci, have been shown to be recessive in the hybrids. The same thing has been shown for 23 recessive mutant genes of simulans, in 23 loci. Allowing for loci represented in both totals, recessives in 70 loci have been shown to be recessive in the hybrids as well as in the parent species — and not a single exception to the rule has been found. Five sex-linked recessives of simulans have been shown to produce their usual effects in hybrid males; a sixth, bobbed, failed to produce such an effect — but bobbed overlaps wild type even in simulans, and in melanogaster it never appears in males, so that this is not a real exception to the rule that mutant characters behave in the hybrids as they do in the pure species. Sixteen dominants have also been tested in the hybrids. All except Lobe (of melanogaster) showed their usual character. Lobe was recessive, but here again we are not dealing with a real exception, for even in melanogaster heterozygous Lobe is not always distinguishable from wild-type, and the character (a small eye) is opposed in the hybrids to an eye larger than that of melanogaster, derived from the simulans parent. Bar and Lobe-2 also show the effects of this larger eye, in that their effects are much decreased, though still obvious, in hybrids.

The data just presented show that each species carries the dominant wild-type allelomorphs for the mutant genes of the other species as well as for its own mutant genes. It thus becomes possible to apply a test to suspected corresponding genes. If scarlet melanogaster, when crossed to scarlet simulans, gives scarlet hybrids, then it follows that neither stock carried the dominant wild-type allelomorph of the mutant gene present in the other stock; therefore the two mutant genes represent modifications of the same wild-type gene, or at least of wild-type genes with correspondings effects on development. Within a given species such a result has been found to be a certain index of allelomorphism, so it may safely be assumed that the two scarlets have the usual properties of allelomorphs, except perhaps that of acting as mates at maturation.

Extensive tests of this sort have been made between the mutant types of simulans and those types of melanogaster that resemble them. In 17 cases the hybrids have shown the mutant characters in question;

all the other characters tested have given clear negative results. Figure 61 shows the maps of the chromosomes of the two species.. The terminal loci of all chromosomes are indicated; otherwise only corresponding genes are shown, and these are connected by dotted lines. It is obvious at once that the two X chromosomes agree as to sequence of genes, and that even the crossover values are not strikingly different

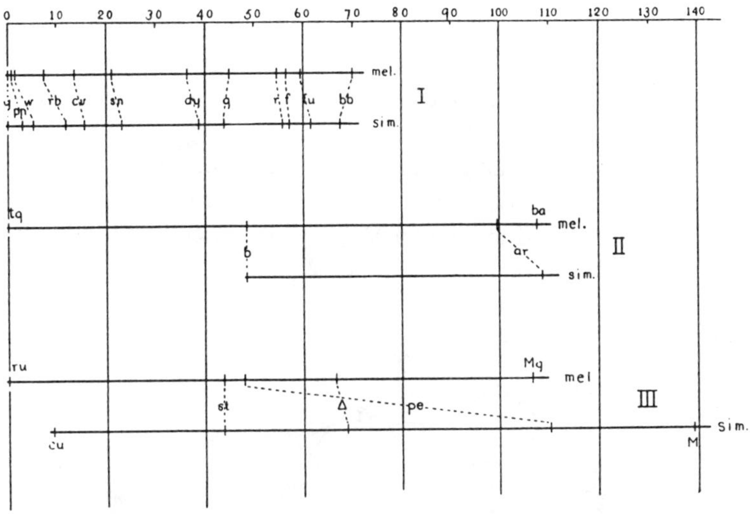

FIG. 61. Chromosome maps showing loci of corresponding genes of D. melanogaster and D. simulans.

in any case. Clearly the two chromosomes are extremely similar in make-up. There is much less information for chromosome II, but there is no indication of any striking specific difference. In chromosome III, however, the sequence and the crossover values are both very clearly and strikingly different. It may be noted here that scarlet and peach are the most thoroughly established as corresponding genes of any of the 17 here discussed, so that the discrepancy is certainly a real one. The bearing of this result on the study of parallel mutations in species that cannot be crossed, will be discussed below.

Non-disjunction and gynandromorphs

Non-disjunction of the sex-chromosomes occurs in simulans, and so far as studied is of the same nature as in melanogaster. The percentage of secondary exceptions was 2.9 in the only adequate series of data (STURTEVANT '21*a*).

Gynandromorphs also occur. Seven were described by STURTEVANT ('21*a*), and numerous others have been found more recently. Most of these can be explained on the assumption that one X-chromosome was eliminated at a cleavage division (as in most of the melanogaster gynandromorphs). One specimen, however, was a mosaic both for sex and for mutant characters in chromosome III. Analysis indicates that in this case two sperm must have taken part in the production of the embryo, so the mosaic is to be referred either to a double-nucleus egg or to polyspermy in a single-nucleus egg.

Several mosaics probably due to somatic mutation have been found. Two are on record (STURTEVANT '21*a*); a third was the original specimen of the sex-linked mutant, dusky. The dusky character was pre-

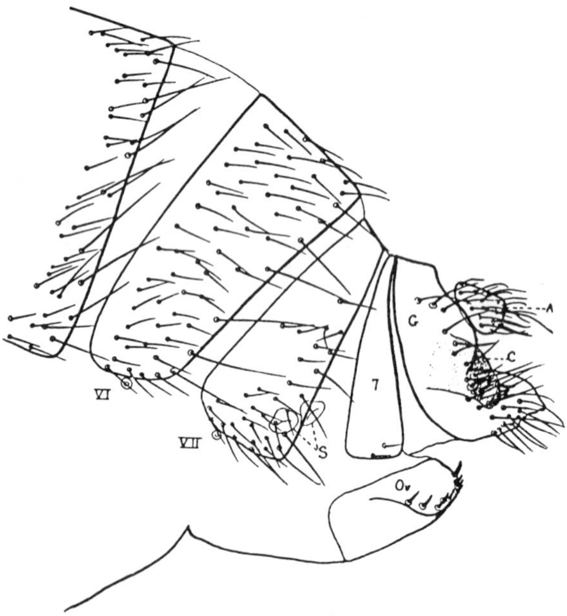

FIG. 62. Abdomen of intersexual Drosophila simulans.

sent in one wing but not in the other; and some, but not all, the X-bearing sperm carried dusky. This case is referred to in the section on Mutation, and has also been referred to by HUETTNER ('23) in connection with the cell-lineage of the primordial germ-cells.

Intersex

The most important single mutant type of simulans is the second chromosome recessive "intersex." This mutant gene, when homozygous, turns females into intersexes and sterilizes males. The intersexes (Fig. 62) are of a definite and constant type, not at all variable like the triploid intersexes of melanogaster. They resemble females in having no sex-combs on the fore tarsi, in having seven distinct abdominal tergites and in the presence of chitinized spermathecae and egg-guides. Male characters are lateral anal plates, claspers, black tip to abdomen, and "first genital tergite". The gonads are rudimentary and the intersexes are sterile. By the use of the sex-linked character yellow it has been possible to show that the male parts of these intersexes carry two X-chromosomes. It seems clear that we are here dealing with a mutation in one of the autosomal genes concerned in the differentiation of sexual characters, such as the data from triploid races of melanogaster show must exist (see section on Sex-determination).

CHAPTER XX

Genetics of other species of Drosophila

Drosophila busckii Coquillett

Warren ('17, '20) has reported three mutations. Red and chocolate are recessive autosomal eye-colors that apparently lie in a single chromosome pair. The descriptions suggest that they resemble scarlet and sepia of melanogaster, respectively. Abnormal is a sex-limited autosomal type, not linked to red or to chocolate. The character, which appears only in females, consists in the absence of two median spots normally present on the fifth abdominal sclerite. The females heterozygous for abnormal may or may not show the character, depending on temperature (low temperature increases the number of abnormals, high decreases it) or modifying genes. Warren has also carried out selection experiments on the spotting of the fifth abdominal sclerite in males, that are normally variable for this character. Selection was effective in isolating "high" and "low" lines, and these lines remained distinct after selection was discontinued. Crosses were made, but no factorial analysis of the variability was completed. There is however a strong probability that a sex-linked modifier was present in one case. The spotting of the males was also affected by temperature.

Drosophila immigrans Sturtevant — *see p. 204.*

Drosophila similis Williston

Metz ('16*b*) described a non-sex-linked recessive eye-color called chocolate. The character apparently resembled sepia of melanogaster and chocolate of busckii.

Drosophila virilis Sturtevant

In November, 1913, a single pair of this species was bred from a pineapple exposed at Columbia University, New York City. These two specimens (one may guess that they were themselves brother and sister) were mated together, and were subsequently preserved, along with a number of their immediate offspring. Until 1919 no other live material of this species was observed. The descendants of the original pair produced, in this time, 19 mutant types in the cultures of C. W. Metz; and these included 3 dominant characters, none of which are to be seen on the original pair. The evidence is thus conclusive that we are dealing, in this case, with actual new germinal changes, rather than with recombinations of pre-existing genes.

Stocks of virilis have since been obtained from Terre Haute, Ind. (R. R. Hyde) and from Los Angeles, Calif. (C. V. Beers); and these have been used in the later experiments of Metz and others. There are no other certain records for the species, which is, in nature, one of the rarest of all the American Drosophilinae.

The earlier papers on the genetics of virilis (Metz and Metz '15; Metz '16*b*, '16*c*, '18, '20*a*, '20*b*; Weinstein '20) have been summarized, together with much new material, by Metz, Moses, and Mason ('23). Further data have been published by Weinstein ('22) and Metz ('23).

Six linkage groups have been found; the mutant genes recorded are summarized in Table 12. We are also indebted to Dr. Metz for some new data included in this table; Dr. Metz's published account will probably appear as soon as the present account.

TABLE 12

Chromosome I

Gene	character affected	locus
sepia	eye color	0
yellow	body color	4.7
scute	bristles	5.4
frayed	abdomen	6
crossveinless	wing veins	24
vermilion	eye color	24.5
vesiculated	wings	26
oblique	wings	36
singed	bristles	42
diminutive	wings	59
wee	wings	60
hairy	eyes	62
magenta	eye color	67
forked	bristles	71
prune, white	eye color	80
Triangle	wing veins	81
short, cut	wing veins	86
glazed, rugose, wax	eye surface	102
droop	wings	109

Chromosome II

Gene	character affected	locus
confluent	wing veins	0
broken	wing vein	42
concave	bristle, wings	65
double	wing veins	

Chromosome III

Gene	character affected	locus
Scaly	eyes	0
spread	wings	20
hunch	thorax	30
telescoped	thorax	48
Garnet	eye color	91
steel	eyes	

Chromosome IV

Gene	character affected	locus
acute	wings	0
pinched	wing veins	0
hump	thorax	0

Chromosome V

Gene	character affected	locus
fused	wing veins	0
Branched	wing veins	23
interrupted	wing veins	26
approximated	wing veins	57
ruffled	hairs, bristles	86

Chromosome VI

Gene	character affected	locus
net	bristle, wings	0
bent	wings, eyes	0

Not Located

Gene	character affected	locus
vestigial	wings	
Spread	wings	
Minus	bristles	
capsule	wings	
spine	wings	
extra	wing veins	

Weinstein ('20, '22) has reported on several special phenomena connected with the behavior of the sex-chromosomes of virilis. Non-disjunction occurs, and 1.3 per cent of secondary exceptions were found. The XO males are probably sterile, as in melanogaster. A gynandromorph is recorded, that probably arose through elimination of the maternal X. The coincidence values are about the same as those found in the X of melanogaster, and the data indicate that the values rise to a maximum and then decrease again as the length of interval studied is increased.

Metz ('23) has found that low temperature intensifies some of the peculiarities of the "bent" mutant type, and that similar results are observable when the melanogaster type, of the same name, is subjected to low temperatures.

Drosophila funebris Fabricius

Only three sex-linked genes have been studied in this species, yet the map of the X-chromosome is 70 units long without correction for unobservable double crossing over. Notch, a dominant that is lethal in males and that closely resembles the characters of the same name in melanogaster and obscura, was described by STURTEVANT ('18*d*). Forked, a recessive bristle character closely similar to the forked especially of virilis, was described by STURTEVANT ('21*a*). Bubble, a recessive wing-character resembling inflated of melanogaster and vesiculated of simulans or virilis, was recorded but not described in the latter paper. The available linkage data for these three loci follow:

TABLE 13

Loci	*Totals*	*Crossovers*	*Percentage*
N f	5951	1684	28.3
N bb	1185	546	46.1
f bb	1101	471	42.3

That the sequence is really N f bb has been shown by three different types of experiments involving all three loci at once. These experiments may be summarized as follows:

TABLE 14

Nature of cross	Non-crossovers	Single crossovers		Double crossovers	Total
		Region 1	Region 2		
N × f bb	85	35	52	23	195
N f × bb	240	91	193	71	595
N f bb × +	80	24	42	11	157
Total	405	150	287	105	947

These data indicate that crossing over in the two regions is practically independent — coincidence is 100.4.

Non-disjunction and gynandromorphs have both been carefully searched for in funebris, but without success, in spite of the fact that forked is an unusually favorable character for use in detecting such irregularities.

Four autosomal mutant genes are known in funebris. MOHR and STURTEVANT ('19) recorded a sex-limited semi-lethal, abnormal abdomen, affecting females almost exclusively. Both the abnormal abdomen and the lethal effect were strongly affected by unanalyzed environmental influences.

The other three autosomal mutant characters are:

1) *Hairless*, a dominant closely resembling the Hairless of melanogaster;

2) *Tiny-bristle*, a recessive that causes some or all the thoracic bristles to be much reduced in size;

3) *Pointed*, a recessive that makes the wing slightly smaller and more pointed at the apex than in typical funebris. These last three have been shown to be independently inherited.

Drosophila hydei Sturtevant

The data on this species have been published by HYDE ('15*a*, '22) and CLAUSEN ('23). The species is extremely like D. repleta in appearance, and was not distinguished from it and given a name until 1921 (STURTEVANT '21). Accordingly in the literature up to that time it appears under the name of repleta.

CLAUSEN has described two recessive sex-linked eye-colors, called white and vermilion, that are closely similar to the types of the same names in melanogaster and other species. They give 9.3 % crossing over—distinctly less than the similar types in melanogaster, but very nearly the same as those in obscura.

HYDE has described two recessive autosomal mutant types: scarlet eye, resembling the character of the same name in other species; and variable eye, a small eye that is variable in size, being smallest and most easily separable from wild-type when the flies are reared under warm and dry conditions. The two genes are linked; HYDE's data indicate that they do not cross over in males, but show about 25 % of crossing-over in females.

CLAUSEN has also described a mutant character called "bobbed", that resembles the characters of the same name in melanogaster and simulans. The inheritance of this character has not been reported.

Drosophila repleta Wollaston

STURTEVANT ('15*c*) reported a sex-linked recessive, called "light", found in nature. This type has the dark thoracic spots, characteristic of typical repleta, much reduced in size. It is figured by MORGAN ('16, page 76); the normal there shown for comparison is too light for typical specimens — it was most probably a heterozygous female. In the original paper the light type was reported as occurring in numerous localities; but, as pointed out later by STURTEVANT ('21*b*, page 14), the classification was based on material that also included hydei and mulleri, which had not then been distinguished from repleta. Careful reexamination of the available pinned material (from many parts of the United States, the American tropics, and Europe) has disclosed light repleta only from New York City and from Los Angeles, Calif. (C.V. BEERS). In the original paper a male from Arkansas was reported to have been tested genetically and to have bred as a light, but the later evidence makes it seem possible that that result may in reality have been due to non-virginity of the females to which the Arkansas male was mated. HYDE ('22) records the light type from Baltimore, but this record requires confirmation by comparison with New York lights. There is an undoubted light specimen (a male) collected in the Columbia laboratory in February, 1913. In October, 1914, STURTEVANT ('15*c*) reported that 17 % of the wild repleta taken in New York City were light. The species has not been bred in the Columbia laboratory since 1916, but still maintains itself there. The following counts of wild material indicate that the percentage of lights is not changing:

TABLE 15. — DROSOPHILA REPLETA TAKEN IN THE COLUMBIA LABORATORY

	♀♀		♂♂	
	Dark	Light	Dark	Light
February, 1923	47	0	53	8
November, 1923	42	4	57	11
February, 1924	32	1	47	7
May, 1924	45	1	34	7

STURTEVANT ('21*b*) has reported "two other mutations in this species, but neither of them was favorable for extensive study." One of these was a modifier of the light type, of such a nature as to make the females heterozygous for light scarcely distinguishable from homozygous light. The other was a reduction in number of scutellar bristles, apparently sex-linked and crossing-over freely with light.

The eye-color mutation reported by HYDE ('15*a*) was really in D. hydei, q. v.

Drosophila obscura Fallén

This species was first studied by METZ ('16*b*), who reported three mutations in it. LANCEFIELD ('19, '22), especially in his second paper, gives an account of a very extensive series of sex-linked mutations. The autosomal mutations studied by LANCEFIELD are tabulated (named by symbol only and not described) by chromosomes by MORGAN ('22*b*, p. 182).

The accompanying table (16) shows the arrangement of the sex-linked genes. A discussion of the possible parallelism between this series of loci and those found in the other species will be presented after the various species have been described individually. It will be observed that this is the longest map recorded in the present volume—the longest, in fact, yet met with in the study of any organism.

Four autosomal groups of genes are here reported, with 4, 6, 1, and 2 recognized loci respectively. These, together with the sex-linked group, make the number of linkage groups 5 — which is also the haploid number of chromosomes in obscura. Professor LANCEFIELD has furnished a list of these autosomal genes, which is here included in table 16.

TABLE 16. — MUTANT CHARACTERS OF DROSOPHILA OBSCURA

Chromosome I

Name	locus	character affected
Pointed	0	wings
glazed	4	eye surface
beaded	20	wings
scutellar	61.7	bristles
yellow	62	body color
Notch	67	wings, etc.
white, eosin	68	eye color
vermilion	72	eye color

Name	locus	character affected
singed } polished }	74	bristles, etc.
miniature	74	wings
reduced	85	bristles
Broken	90	wing veins
rough	96	eye surface
fused	104	wing veins
compressed	109	thorax, eyes, etc.
ascute	148.5	thorax
slender	161.5	bristles
short	169.5	wing veins
stubby	170	bristles

Chromocome II

Name	locus	character affected
confluent	?	wing veins, etc
spread	?	wings
rough II	?	eye surface
bithorax	?	balancers

Chromosome III

Name	locus	character affected
scarlet	0	eye color
Trimmed	0	wings
small	30	bristles
Minute	30	bristles
gap	26 or 34	wing veins
lethal	?	life

Chromosome IV

Name	locus	character affected
Knobby	0	eye shape

Chromosome V

Name	locus	character affected
Achaete	?	bristles and hairs
Missing	?	bristles

Drosophila affinis Sturtevant

HYDE ('15*b*) reported a jaunty wing, inherited as a non-sex-linked recessive. The species was then undescribed, and HYDE used the name "confusa" for it.

Further unpublished data are available on an autosomal recessive eye-color, closely resembling the peach of melanogaster or of simulans. A curious sex-ratio has also been studied. The stock containing this modification has been lost, so the analysis cannot be completed, but the data obtained indicate that a great deficiency of sons was obtained from certain males, regardless of the source of the females to which such males were mated. The few males obtained from such matings were, usually at least, normal in behavior; but some of the sons of the females from such anomalous cultures again gave very few sons.

Drosophila caribbea Sturtevant

STURTEVANT ('21*b*) recorded an autosomal recessive, curved wing. The character resembled the curved of melanogaster or of simulans. No detailed account of it has been published, nor does it seem worth while to present such an account now.

Drosophila willistoni Sturtevant

The early data on the chromosomes of this species were published (METZ '14*b*) under the earlier (but preoccupied) name of Drosophila pallida WILLISTON. The chromosome group of the female is the same as that of the melanogaster female, except that here the small spherical fourth chromosome pair appears to be absent. LANCEFIELD and METZ ('21) showed that this similarity is only apparent. They studied the chromosome groups of non-disjunctional females, identified by means of sex-linked mutant characters, and were able to show that the X-chromosome is V-shaped in willistoni, while the rod-shaped chromosome that resembles the X of melanogaster is an autosome.

LANCEFIELD and METZ ('22) described 28 sex-linked mutant characters, and published linkage data and a map for the X-chromosome. METZ and FERRY ('23) reported two more sex-linked mutant characters; and FERRY, LANCEFIELD, and METZ ('23) published descriptions and figures of some new sex-linked types and of all the autosomal mutant characters then known in willistoni. The recorded mutant characters, with a small amount of additional data (furnished us by Dr. METZ) bearing chiefly on their loci within the chromosomes, are listed in Table 17. It will be observed that only three groups of linked genes have been found, and it was noted above that only three pairs of chromosomes are present.

TABLE 17. — MUTANT CHARACTERS OF D. WILLISTONI

Chromosome I

Mutant	locus	Character
rosette	0	bristles
nicked	8	wings
morula	14	eye surface
short, etc.	16	wing-veins
stump	24	wing-veins
pale	28?	bristle-color
rimmed	33	wings
deformed	34	eyes, etc.
triple	34.5	wing-veins
rough	35	eyes
beaded	38	wings
peach	40	eye-color
scute, etc.	41.9±	bristles
yellow	42	body-color
reduced	43	bristles
crossveinless	44	wing-veins
vermilion	44.5	eye-color

Mutant	locus	Character
square	53	wings
forked, etc.	54	bristles
tiny	56	bristles
bent	74	wings
small-bristle	79	bristles
orange	83	eye-color
stubby	84	bristles
interrupted	?	wing-veins
bald	?	bristles
abbreviated	?	wing-veins
broken	?	wing-veins
Chromosome II		
circular	0	wings
blunt	27	wings
spread	33	wings
approximated	39	wing-veins
Knotted	81	wing-veins
Clipped	?	wings
apterous	?	wings

Mutant	locus	character
ragged	?	wings
red	?	eye-color
balloon	?	wings
Chromosome III		
tardigrade	?	legs
combined	?	thorax etc.
glossy	?	eyes
incomplete	?	wing-veins
blistered	?	wings, etc.
rounded	?	wings
absent	?	wing-veins
granulated	?	wings, etc.
Chromosome not Reported		
branched	?	wing-veins
lacking	?	bristles
cone	?	tarsi
jaunty	?	wings

Genetic Studies on Diptera other than Drosophila

Definite Mendelian characters have so far been recorded in only two Diptera outside the genus Drosophila.

PAINTER ('17, Amer. Nat. 51: 306—308) has reported an autosomal recessive mutant character in *Piophila casei* LINNÉ. The mutant type somewhat resembled the balloon wing of Drosophila melanogaster; like that type it was quite variable in appearance. Piophila is an Acalypterate; i. e., it belongs in the same large group of subfamilies as do the Drosophilinae. It has the same type of chromosome group as most Muscoids (see above).

The other species that has yielded a definite mutation is *Psychoda alternata* SAY, one of the Psychodidae, belonging at the other end of the order Diptera from the Drosophilines. In this form TURNER ('23, Amer. Nat. 57; 545—558) has reported an autosomal recessive, white eye-color. The chromosome group of Psychoda has not been reported.

Various authors have recorded results of breeding experiments with the Calypterate genera Lucilia, Calliphora, and Cynomyia. There is nothing in this literature of interest for a comparison with the work on Drosophila. The following list of references may be useful as a matter of record.

CUÉNOT; 1889. Bull. Sci. France Belg. 32 (sex ratio in Calliphora, Lucilia, etc.).

TOTHILL; 1913. Ann. Ent. Soc. Amer. 6; 241—256 (bristle number in Lucilia).

WHITING; 1913. Ann. Ent. Soc. Amer. 6; 257—267 (bristle number in Lucilia and Calliphora).

WHITING; 1914. Amer. Nat. 48; 339—355 (bristle number in Lucilia)

WHITING; 1914. Biol. Bull. 26; 184—194 (pupation and color in Lucilia, Cynomyia, and Calliphora).

VALENTI; 1915. Bios 2; 265—297 (sex ratio in Calliphora).

ENRIQUES; 1919. Rivista Biol. 1; 72—81 (non-viable larvae in Calliphora).

Drosophila immigrans Sturtevant.

This species, overlooked when the manuscript was written, has been studied by METZ and METZ ('15) and by STURTEVANT ('21b). A sex-linked small wing was found, and four autosomal genes. Of the latter, one was a small-bristle character, and the others were venation changes.

CHAPTER XXI

Parallel and corresponding genes

Related species must have many genes in common—they are in fact similar only because of this group of common genes. In a given species certain mutations recur, in some cases rather frequently. It is therefore probable that in separate but related species the same mutation will sometimes occur independently. Most of the work on species of Drosophila other than melanogaster has been based on these considerations, and on the hope that if identical mutations did occur it would be possible to identify them as such.

As pointed out above in discussing Drosophila simulans, the usual test for allelomorphism within a species is that of crossing recessives and observing whether or not the hybrid shows the recessive character. It seems advisable to restrict the term "allelomorphic" to genes that act as mates at segregation. In the present discussion we shall use "corresponding" to designate genes that fulfill the above conditions but may occupy separate loci. These are, evidently, genes that play corresponding roles in development. The term "parallel" we shall use for genes suspected of being corresponding because of a similarity in the characters produced. In the case of the Drosophila species this test can be applied only between melanogaster and simulans, since no other interspecific hybrids have been obtained.

If the sex-linked mutants of obscura, willistoni, or virilis are compared with the mutant types of melanogaster it soon becomes evident that many more of them show close resemblances to the sex-linked mutants of that species than to its autosomal mutants. It thus seems probable that a high proportion of these resemblances are due to an actual genetic similarity—that in many cases these "parallel" mutations really depend on corresponding genes. This conclusion is strongly supported by the high percentage of known corresponding mutations in simulans.

When we compare the sex-linked parallels in different species as to sequence, we find that the maps are quite diverse. In virilis, for example, there are 10 good parallels to melanogaster mutations. In melanogaster the sequence is y, sc, w, cv, lz, sn, v, m, dy, f; in virilis it is y, sc, cv, v, sn (or f), m (or dy), dy (or m), f (or sn), w, lz. If we suppose y is a corresponding gene, there are 7 loci of the 10 in virilis in the expected sequence; if we take was the corresponding gene and invert the virilis sequence, there are 5; and neither scheme utilizes v and lz, which are among the most convincing of them all when considered apart from sequence. Even the series of 7 in the expected sequence is apparently not more than might easily result from mere chance.

The data from chromosome III of simulans show that a modification of the sequence of corresponding genes may occur, and it seems probable that this has occurred in the X-chromosomes of obscura, virilis, and willistoni (LANCEFIELD '22; METZ, MOSES and MASON '23). It is, however, not possible to state definitely that any given difference in sequence of parallel types is due to a rearrangement of parts of a chromosome; for it is not possible to be certain that any given parallels are really due to corresponding genes.

The nature of the difficulties encountered in identifying corresponding genes has been discussed in detail by STURTEVANT ('21*a*). We need point out here only that, even within a species, similarity of mutant characters has been found a very unreliable guide as to community of origin; and between separate species this difficulty is magnified because of the different reaction-system in which the parallel mutations produce their effects.

There follows a brief discussion of each of the cases of parallelism between sex-linked characters.

Yellow—melanogaster (0), simulans (0), virilis (4.7), obscura (62), willistoni (42). This is one of the frequently recurring types in melanogaster, and has probably occurred more than once in simulans and in virilis. The appearance of the mutant flies in the different species is as similar as would be expected in view of the different body-colors shown by the various wild-types.

Two other similar body-colors have been found in melanogaster, but neither of these resembled the yellow of other species as closely as does yellow melanogaster.

The position of the locus is terminal in melanogaster and simulans, subterminal in virilis; but in obscura and willistoni it is median or submedian. The first three species have rod-shaped X-chromosomes with terminal spindle-fiber attachments, while the last two have V-shaped ones with median attachment. But the significance of these relations has become extremely doubtful with the discovery that yellow in melanogaster lies at the free end of the X, not at the end attached to the spindle-fiber.

Scute — melanogaster (0.1), virilis (5.4), obscura (62-), willistoni (42-). This has recurred in melanogaster, where it is probable that at least two different mutant allelomorphs occur. In willistoni three allelomorphs are known. In all the types here included the scutellar bristles are reduced in number, and in most of them some of the head bristles are also missing. Somewhat similar characters are known in other loci, but none that is very much like these. It is noteworthy that in all four species the locus of scute is close to that of yellow.

Prune — melanogaster (1.0), simulans (2.9). These two have been shown to be corresponding types, by examination of the hybrids. Prune has occurred only once in each species. A character of the same name in virilis, somewhat similar to this, is an allelomorph of white and so probably does not correspond to prune melanogaster. Sepia of virilis may also be compared to prune. Eye-colors of the same general type as this and the "purples" and "pinks" are so frequent in most species that they do not furnish favorable material for identifying parallel mutations, though a careful study of their histology and of their dilution effects on each other would probably make them available for this purpose.

White — melanogaster (1.5), obscura (68), virilis (80), hydei (0). In melanogaster white is one of the most frequent mutations, and it has a series of 10 multiple allelomorphs. In obscura it has occurred twice, and also has an allelomorph, eosin, that is quite similar to the white allelomorph of the same name in melanogaster. In obscura the white female is sterile, which is not the case in any of the other mutants of this type. In virilis also there is an allelomorph, prune, perhaps to be compared to some of the darker allelomorphs of white in melanogaster.

Notch — melanogaster (3), obscura (67), funebris (0). In melanogaster notch has occurred about 20 times, in obscura twice. In melanogaster it has been shown by MOHR ('19, '23*d*) to be a deficiency for several loci, and for different ones in different cases. In obscura and funebris there is no proof of a deficiency, but that is the most probable interpretation.

The character is quite complex in all three species, and the peculiarities agree in considerable detail.

Ruby — melanogaster (7.5), simulans (11.9). This locus has mutated three times in melanogaster, twice in simulans. The character is practically indentical in appearance with garnet (see below), and with several autosomal characters. Among the sex-linked characters it may be compared with sepia (0) and magenta (67) of virilis and with peach (40) and orange (83) of willistoni. For the reasons stated in the discussion of prune, these comparisons are not very satisfactory.

Crossveinless — melanogaster (13.7), simulans (15), virilis (24), willistoni (44). The mutation has been found once in each of the species named. The characters agree in detail, but there are others known that resemble them somewhat, e. g., the two third-chromosome recessives "incomplete" and "absent" of willistoni (FERRY, LANCEFIELD, and METZ 1923).

Lozenge — melanogaster (27.7), virilis (102), obscura (4), willistoni (14). The characters in question are called lozenge in melanogaster, rugose (with allelomorphs glazed and wax) in virilis, glazed in obscura, morula in willistoni. All the types have abnormal surfaces to the compound eyes, often suggesting the presence of moisture. In melanogaster and in virilis three allelomorphs are known that differ in the extent of eye modification, and in each species two of these have sterile females.

Singed and forked — melanogaster (sn, 21.0; f, 56.5), simulans (sn, 23; f, 58) virilis (sn, 42; f, 71), funebris (f, 28), obscura (sn, 82), willistoni (f, 54; stubby, 84). These two mutant types are considered together, since the characters are closely similar. The symbols used refer only to the accepted names, and are not to be taken as identifications of corresponding genes except in the case of melanogaster and simulans. In melanogaster each has recurred; forked is represented by 3 different allelomorphs, singed by 3. As judged by structure no consistent difference is to be found between the two series (in melanogaster and simulans) but a difference does occur in that singed females are sterile (though singed-2 females are fertile), while sterility has not been found associated with any forked allelomorph. The females are sterile in both forked allelomorphs (but not in stubby) of willistoni, and in singed of obscura. None of the other types under discussion has sterile females.

A detailed comparison of the types is given by METZ, MOSES, and MASON ('23, pp. 59–61).

Cut — melanogaster (20.0), obscura (20), willistoni (38), (known as beaded in obscura and willistoni). This is one of the frequently recurring mutations in melanogaster, where several different allelomorphs are known. The character is one that is approximated by mutant types dependent on several other genes, so that the case is not a good one for establishing parallelism.

Vermilion — melanogaster (33.0), virilis (24.5), hydei (9), obscura (14), willistoni (2.5). The characters concerned are as similar as would be expected in view of the differences in wild-type eye-color in the several species. The mutation is one of the most frequently recurring ones in melanogaster. The character is, however, nearly duplicated by the third-chromosome type scarlet (of melanogaster, simulans, and obscura), and by the melanogaster types cardinal (chromosome III) and cinnabar (chromosome II).

Miniature and dusky — melanogaster (m, 36.1; dy, 36.2), simulans (dy, 39), virilis (diminutive, 59; wee, 60), obscura (m, 82). These characters are as difficult to distinguish as are forked and singed, and are therefore here treated in the same way as were those. Each has recurred in melanogaster, where dusky is represented by two allelomorphs—neither of them as much like miniature as is the corresponding dusky of simulans.

Garnet — melanogaster (44.4), simulans (44). Two allelomorphs are known in melanogaster, and the corresponding simulans type is intermediate between the two. For possible parallels in other species see discussion under ruby.

Inflated — melanogaster (53), funebris (71). The character, known as bubble in funebris, belongs to a rather common type of modification. It was first compared by WEINSTEIN ('20) to vesiculated, to which it bears some resemblance.

Rudimentary — melanogaster (54.5), simulans (56), willistoni (53). The resemblance of rudimentary (known to be corresponding in melanogaster and simulans) to square, willistoni, is not very striking.

Fused — melanogaster (59.0), simulans (63). Two mutant allelomorphs are known in melanogaster. The fused females are almost entirely sterile in melanogaster (LYNCH '19), but are fertile in simulans. Crosses have shown that these are really corresponding genes.

Bobbed — melanogaster (70.0), simulans (68). These are the terminal loci of their respective maps, and crosses have shown them to be corresponding.

Vesiculated — simulans (21), virilis (26). The character is very poor for linkage purposes as in both species it overlaps wild-type. It is also unsatisfactory as a parallel because it is so similar to numerous other characters

Short — virilis (86), obscura (169.5), willistoni (16). There are three different allelomorphs each in obscura and in willistoni. Aside from this the character (a shortening of the 5th and sometimes of other veins) is of too common occurrence to be altogether satisfactory as a parallel.

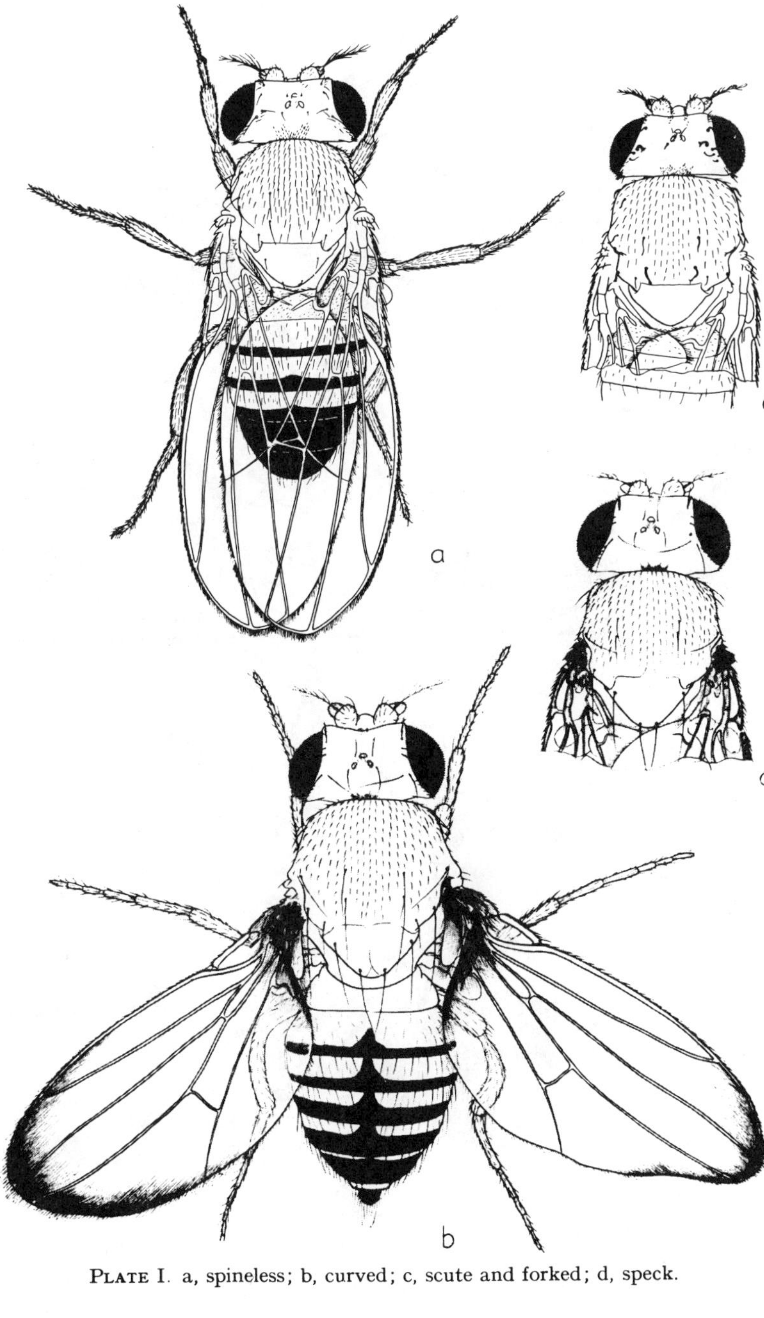

PLATE I. a, spineless; b, curved; c, scute and forked; d, speck.

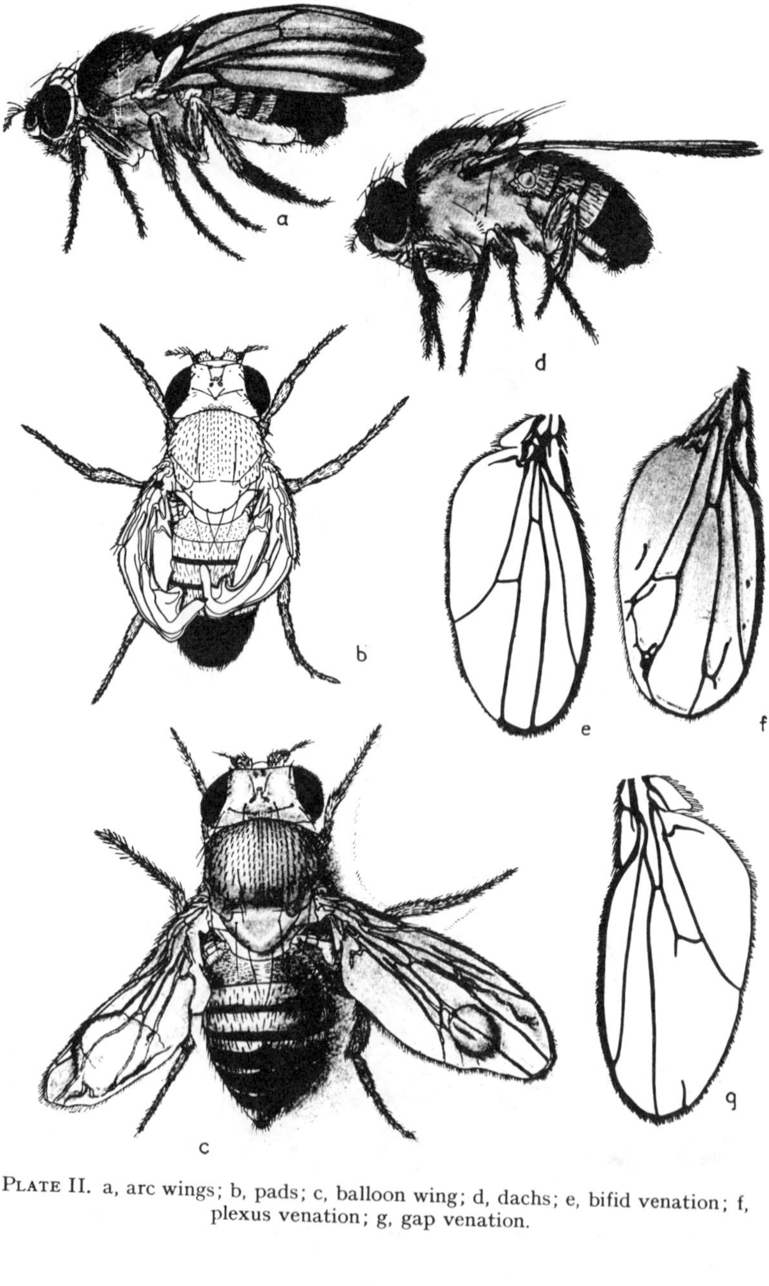

PLATE II. a, arc wings; b, pads; c, balloon wing; d, dachs; e, bifid venation; f, plexus venation; g, gap venation.

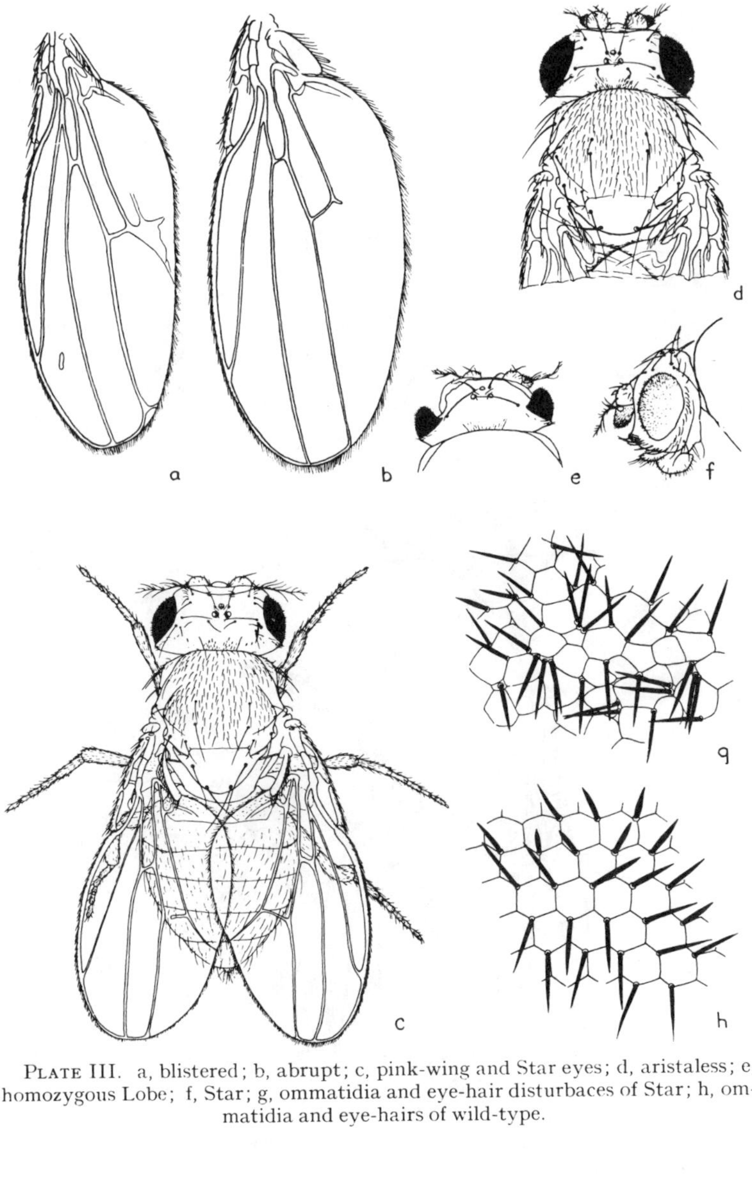

PLATE III. a, blistered; b, abrupt; c, pink-wing and Star eyes; d, aristaless; e, homozygous Lobe; f, Star; g, ommatidia and eye-hair disturbaces of Star; h, ommatidia and eye-hairs of wild-type.

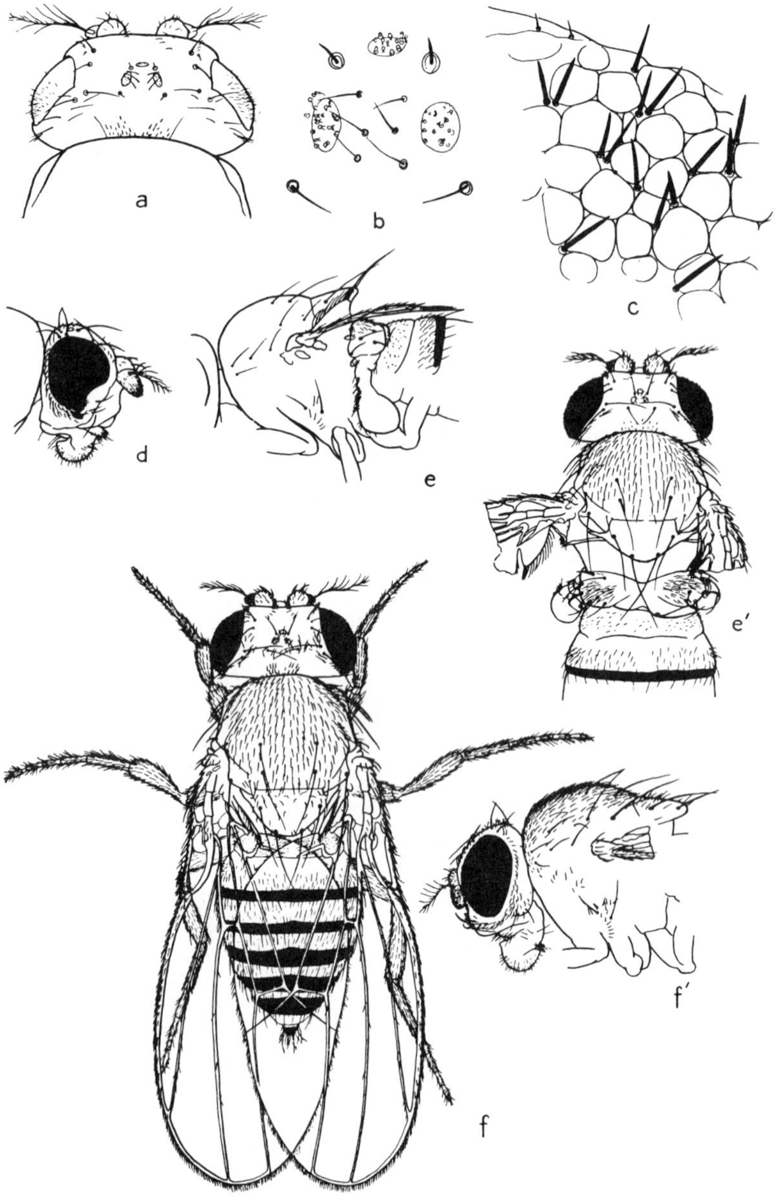

PLATE IV. a, glass eye; b, ocelli of glass; c, ommatidia and eye-hairs of glass; d, kidney eye; e, and e′, side and dorsal views of bithorax-b; f and f′, dorsal and side views of hairy.

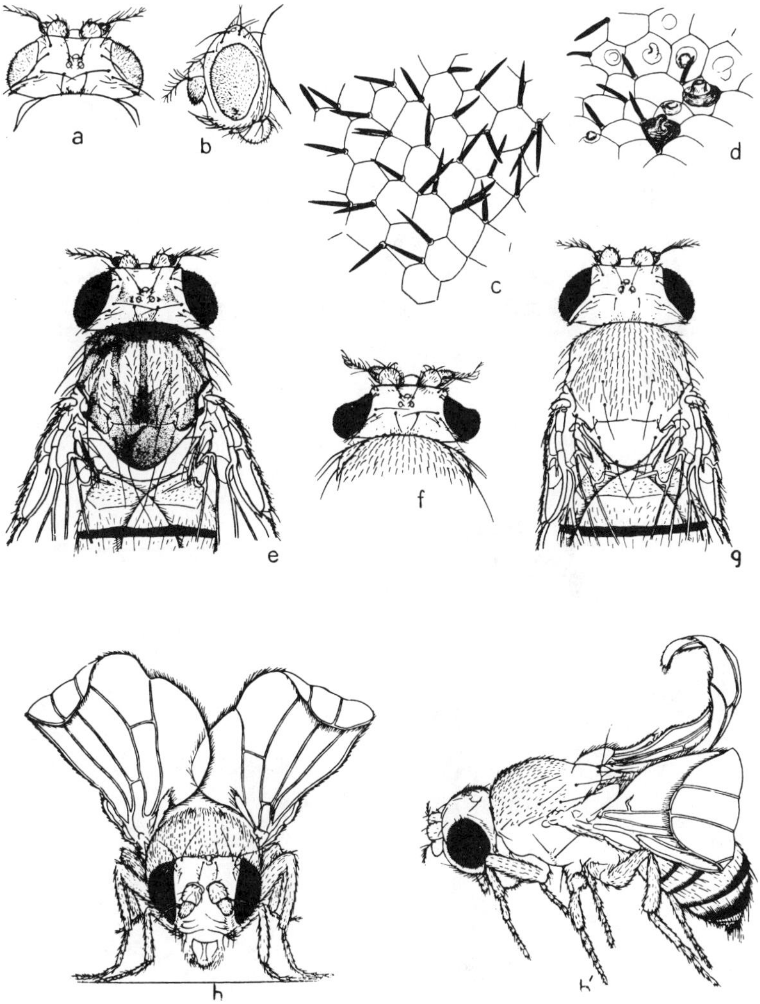

PLATE V. a and b, top and side view of roughoid; c, ommatidia and eye-hairs of less rough part of roughoid eye; d, erupted ommatidia and eye-hairs of smudged part of roughoid eye; e, band; f, deformed; g, minute bristles; h, and h′, front and side views of curled wing.

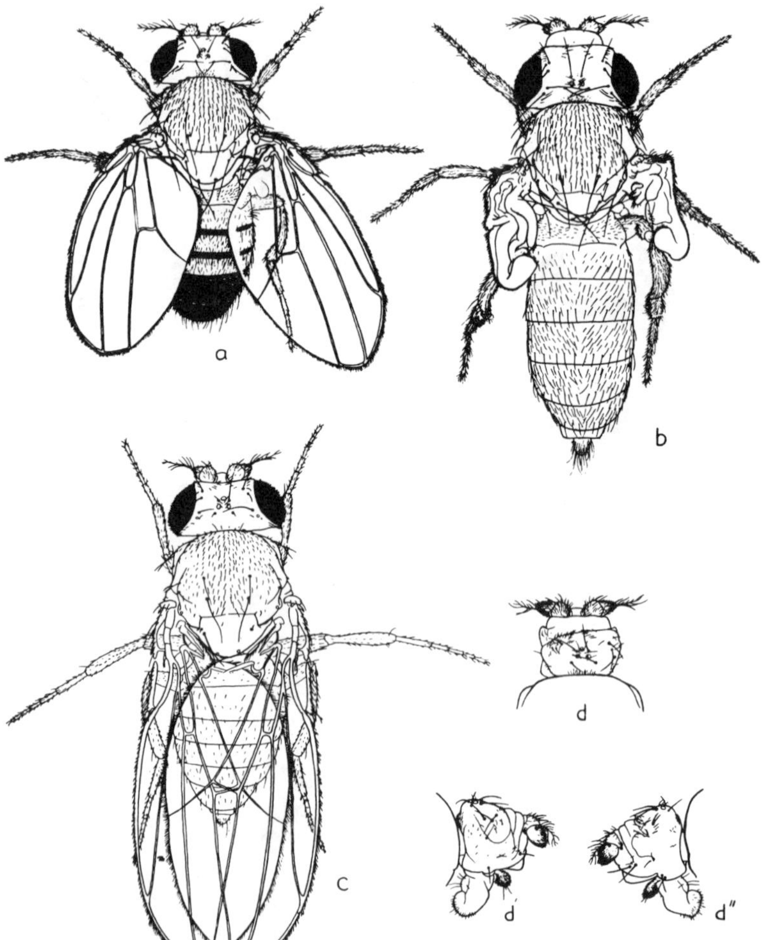

Plate VI. a, bent wings; b, bent, just emerged, shows shortened tarsal joints; c, shaven; d, d′, d″, top and side views of completely eyeless individual from selected stock of eyeless ^{2}.

CHAPTER XXII

List and descriptions of the mutant characters of D. melanogaster

This list is incomplete by the omission of many of the mutants, that are no longer extant, that were little worked with, or that have no longer practical or theoretical importance. The mutants are arranged alphabetically, irrespective of the linkage group. The name of the mutant gene begins with a capital if the character is dominant, and is followed by the symbol. The Roman numerals designate the chromosome and the arabic numerals the locus. The name of the first finder is given, and reference (in parenthesis) to the principal paper in which the mutant has been described.

Earlier references are omitted in case the mutant form has been treated in the Carnegie Publication on the first chromosome (No. 237, M. & B. '16) on the second chromosome (No. 278, B. & M. '19) or on the third chromosome (No. 327, B. & M. '23). References to earlier papers are to be found in those papers and they are also included in the bibliography of this paper. The names of Morgan, Sturtevant, and Bridges are abbreviated to M, S, and B, respectively. The abbreviation (U) means that descriptions of the character and the accounts of its inheritance are either unpublished or very incomplete. In all cases, whether published or unpublished, the descriptions and peculiarities of inheritance are here given briefly with each mutant. Unless noted to the contrary, separability from the wild-type is complete, and viability, fertility and productivity are practically normal. In the maps on page 92 the symbol ! designates a mutant satisfactory in the above respects and also occupying a locus favorable for linkage studies, and furthermore, being a character that can be used with most of the other important mutants without confusion in the classifications. The symbol + designates a mutant somewhat less valuable in some respect. Those unmarked are valuable only in special circumstances.

Abnormal, A. I, 4.5. M. (M. & B. '16, MOHR '23*d*). Abdominal sclerites broken and irregular. Amount of abnormality greatest in early hatches of cultures, and in moist conditions. Extent of abnormality greatest in homozygous ♀, least in het. ♀, intermediate in ♂. Classificationvaries from perfect in some cultures and experiments, to very unreliable in others.

abrupt, ab. II, B. (U). (Plate III, fig. b). Vth-long. vein ends abruptly before reaching margin, usually just beyond posterior cross-vein.

achaete, I. 0.1 ±. WEINSTEIN ('18). Many bristles missing. No basal ring left. Discarded.

amethyst, II. B. (U). Transparent light purplish eye-color. Lost.

antlered, vg^a. II, 67.0. M. (B. & M. '19). (Fig. 15). Allel. of vg, in which wings are extended to form a rt. angle with each other, are nearly full length, but are very narrow strips of the mid-region only. The tip of the longer wings are bent sharply posteriorly. Discarded.

apricot, w^a. I, 1.5. CLAUSEN ('24). Best white allel. for general use. ♂ slightly darker than w^e ♂, ♀ about color of w^e ♂.

apricot dilutor, III. WEINSTEIN (U). Specific recessive dilutor for w^a.

apterous, ap. II, 48.4 ±. WALLACE (B. & M. '19). Wings and balancers reduced to barely perceptible knobs. Viability very poor. Very rarely fertile in either sex. Discarded.

arc, a. II, 99.5. B. (B. & M. '19). (Plate II, fig. a). Wings arched, broad, concave ventrally. Crossveins nearer together.

aristaless, al. II, 3 ±. B. (U). (Plate III, fig. d). Arista reduced in size or entirely absent. Posterior scutellar bristles far apart and strongly divergent.

aristaless-b. III, far to left of D. B. (U). Like aristaless.

ascute, as. III, 43.5. B. (B. & M. '23). Obliteration of furrow between scutellum and thorax. Discarded.

ascutex, ax. I, 28 ±. B. (U). Furrow between scutellum and thorax partly gone. Body-color pale. Black spots at joints in leg. Viability poor. Discarded.

balloon, ba. II, 107.5. M. (B. & M. '19). (Plate II, fig. c). Wings inflated, small, widely extended, discolored, with extra veins. Viability somewhat poor.

band, bn. III, 72.0. M. (B. & M. '23). (Plate V, fig. e). Band of darker pigment across anterior part of the thorax, and an intensification of the trident pattern.

B a r, B. I, 57.0. TICE (M. & B. '16; S. & M. '23; ZELENY '22, '23 etc.; this paper, pages 29—32). (Fig. 13). Eye restricted to narrow vertical bar of 90 ± facets in the ♂, 70 ± in hom. ♀, and 360 ± in het. ♀ as compared with normal numbers of 740 ± for ♂ and 780 ± for ♀.

B e a d e d, Bd. III, 93.8. M. (B. & M. '23). (Fig. 19). Wing margin scalloped. Lethal when hom. Lost.

B e a d e d i n t e n s i f i e r, III. MULLER (B. & M. '23). Lost.

B e a d e x, Bx. I, 59.6. B. (U). Dominant, not lethal when hom. Wing-margins scalloped. Het. form in ♀ not completely separable from wild-type in late hatches.

b e n i g n. III, 25 ±. STARK (B. & M. '23). Black tumor-like growths in larvae, persisting in pupae and adults. Not lethal. Polygenic, main gene a recessive in III.

b e n i g n-b. III. B. (U). Like benign; being studied by STARK.

b e n t, bt. IV, 0.0. MULLER ('14*a*) (Plate VI, figs. a, b). Wings extended and bent backward. Rear-legs knobby. Overlaps wild-type.

b e n t², bt². IV, 0.0. M. (U). More extreme allel. of bt.

b i f i d, bi. I, 6.9. M. (M. & B. '16). (Plate II, fig. e). All long. veins, especially noticable for III and IV, fused at base of wing into bifid stalk. IVth long. often fails to reach margin, with notch in margin. Long. III expands into a small delta where it joins the costal vein.

b i t h o r a x, bx. III, 58.7. B. (B. & M. '23). (Fig. 37). The metathorax, an inconspicuous body-segment behind the scutellum, is changed into a mesothorax in that the balancers become wing-like structures and the bristles are like those of the normal mesothorax. Very variable in amount of change but rarely overlaps the wild-type.

b i t h o r a x-b, bx-b. III, 59.5. B. (B. & M. '23). (Plate IV, figs. e and e′). Like bx, but not allel. Balancers give rounded wing-like structure, not thin as in bx.

b l a c k, b. II, 48.5. M. (B. & M. '19). Body-color black, veins black-bordered.

b l a c k-l e g. III, close to pink at 48. B. (B. & M. '23). Hind legs bowed and blackened. Body-color pallid. Discarded.

b l i s t e r e d, bs. II, 105 ±. B. (B. & M. '19). (Plate III, fig. 1). Wings have small vesicle or plexus of veins at juncture of posterior crossvein and long-V. Wings smaller, darker.

b l o o d, w^b I, 1.5. HYDE (HYDE '16) Transparent pink eye-color that fluctuates from a light yellowish pink (like "eosin" or "apricot") to a deep purplish pink (like "purple"). Allel. of white.

b o b b e d, bb. I, 70.0 B. (U) (MOHR '23*d*). This mutant, (or undistinguished allelomorphs) has reappeared several times. Sex-limited, ♂ entirely normal, ♀ has minute bristles and abdominal bands abnormal at the sides. Class. usually excellent, but in some cultures, especially late, and in mass cultures, the character varies to the extreme of all bobbed ♀'s being wild-type in appearance.

b o r d e r e d. I, 70 ±. B. (U). Wings smaller, slightly extended, veins bordered by darker bands, venation ragged. Viability poor. Class unreliable. Arose in XXY ♀ as mutation in one, but not in both of the X's.

B o w. I. B. (M. & B. '16). Arc wing that overlapped wild-type. Lost.

B r i s t l e d, Br. II, 21 ±. MANN ('23*a*). Extra sterno-pleural bristles. Overlaps wild-type. Lethal when hom.

b r o a d, br. I, 0.6. B. (U). (Fig. 49, right wing). Wing about 60—80% of normal length, broad, with full round tip. Crossveins closer together.

b r o w n, bw. II, 105.0. WAALER ('21). Transparent pinkish brown eye-color.

b r o w n², bw². 105.0. Mrs. S. (U). Allel. of bw, darker. Discarded.

b u f f, w^{bf}. I, 1.5. Safir ('16). Eye-color pale pinkish yellow, lighter than eosin male.

CIA. I, 7 ±. B (U). Recessive crossing-over modifier in X. Changes w-v recomb. % from 31 to 10, and v-f from 23 to 15. Probably one gene in X and several autosomal modifiers.

CI. I. DETLEFSEN. Reduces crossing-over in X to slight amount. Probably two separate CI genes, and several autosomal modifiers.

CI. I. L. V. M. (U). Dominant. Suppresses nearly all crossing-over throughout X. Lethal in ♂.

CIB. I. MULLER (U). Dominant crossover-suppressor for all of X, accompanied by recessive lethal action. It occurred in an X carrying sc t² v sl B

CcyL. II. WARD ('23). Eliminates nearly all crossing-over to left of purple. Associated with Curly.

CcyR. II. WARD ('23). Eliminates nearly all crossing-over to right of purple, except in high temp. (30° ±) when it allows about 5 %. Associated with Curly.

CIIL. II, to left of b. S. (19*a*). Dominant, that eliminates nearly all crossing-over to left of purple. In hom. form crossing-over is normal.

CIIR. II, to right of purple. S. ('19*a*). Dominant, that eliminates nearly all crossing-over to right of purple. In hom. form crossing-over is of normal amount.

CIIS. II. B. (U). Dominant that reduces crossing-over throughout II. Lost.

CII,III. II, to left of black. B. (U). Eliminates nearly all crossing-over to left of black. Reduces crossing-over in left half of III.

CIII. III, near ebony. MULLER (B. & M. '23). Eliminates nearly all crossing-over to right of pink. In hom. form allows more crossing-over than normal.

CIIIG. III. GOWEN (GOWEN and GOWEN '22). Recessive in III that eliminates all crossing-over in all chromosomes.

CIII (high h-st). III. B. (U). Characterized by h-st value of 17—25 and high values for neighboring regions.

CIII (low D-st). III. B. (U). Characterized by lower values from h to ss inclusive.

CIIIM. B. (B. & M. '23). Eliminates practically all crossing-over to right of pink.

CIIIN. B. (U). Eliminates nearly all crossing-over to left of pink.

CIIIOL. B. (U). Eliminates nearly all crossing-over to left of pink. Associated with orange.

CIIIOR. B. (U). Eliminates nearly all crossing-over to right of pink. Associated with orange.

CIIIPL. III, in left limb. Payne (in press). Eliminates nearly all crossing-over to left of pink. Associated with CIIIPR.

CIIIPR. III. PAYNE (in press). Eliminates nearly all crossing-over to right of pink. Associated with CIIIPL.

CIIIL. III. L. V. MORGAN (U). Eliminates nearly all crossing-over to left of pink. Associated with CIIIR.

CIIIR. L. V. MORGAN (U). Eliminates nearly all crossing-over to right of pink. Associated with CIIIL.

CIII, II. III, in right limb. S. (S'19*a*). Acts like CIII, but not allelomorphic to it. Increases crossing-over in middle of II in presence of CIIL,CIIR.

c a r d i n a l, cd. III, 75.7. JOHNSON (B. & M. '23). A vermilion-red eye-color that tends to overlap wild-type, especially when old. Ocelli white.

cherry, w^c. I, 1.5. SAFIR (M. & B. '16). Yellowish pink eye-color. ♂ about color of w^e ♀, ♀ slightly lighter. Allel. of white.

chlorotic, cc. I, 0.+. MOHR ('23) pale yellow body-color. Viability poor. Lost.

chrome, I. B. (M. & B. '16). Yellowish body-color. Classification unreliable. Lost.

chubby. II. B. (U). Larvae, pupae and adults shorter than normal.

cinnabar, cn. II, 37.5. CLAUSEN ('24.). Eye-color like vermilion, but more translucent.

claret, ca. III, 100.7. B. (B. & M. '23). Pink eye-color.

cleft, cf.· I, 65.0. B. (WEINSTEIN '18). (Fig. 22). Wings smaller, somewhat extended, venation ragged. Third long. vein splits off a branch just beyond the anterior crossvein. Viability poor. Males entirely sterile. Lost.

Club, cl. I, 16 ±. M. & B. ('16). (Fig. 35). Wings tend to remain as in pupa. Bristles absent from sternopleura. Lost.

cluboid, III, B. (B. & M. '23). Wings often unexpanded fully. Lost.

complex. II. B. (U). Thin-textured curved wing with extra veins.

compressed, cp. III, 48.5. B. (B. & M. '23). Eye and head compressed. Lost.

compressed-dilapidator. III, 68.5. B. (B. & M. '23). Specific intensifier of compressed. Double form very inviable. Lost.

comma. II. B. (B. & M. '19). Pair of comma-like marks on anterior dorsum of thorax. About 80 % fail to show. Discarded.

Confluent, Cf. II. B. (B. & M. '19). Veins thickened and knotted, especially II opposite ant. cv. and where it joins costa. Lethal when hom. Viability and productivity very poor. Lost.

cream-a. B. ('19*c*). Specific modifier of eosin, making it somewhat lighter. Lost.

cream-b, cr-b. II, 22.5 ±. B. (B. & M. '19). Specific dilutor for eosin. Discarded.

cream-c. II. B. ('19*c*). Specific dilutor of eosin. Discarded.

cream-III. III, 36.5. B. (B. & M. '23). Disproportionate dilutor of eosin. By itself gives a purplish eye-color overlapping normal.

cream-II. II. B. (B. & M. '19). Specific dilutor of eosin, dominant, not lethal when hom. Lost.

crooked, fw^c. I, 38.0. B. (U). An allelomorph of furrowed; the eyes being nearly normal but the scutellar and other bristles crooked. Lost.

c r o s s v e i n l e s s, cv. I, 13.7. B. ('20*b*). Posterior crossvein absent, anterior crossvein absent or incomplete.

c r u m p l e d, cm. III, 93.0. B. (B. & M. '23). Wings crumpled. Posterior scutellars divergent and stubby.

c u r l e d, cu. III, 50.0. M. (B. & M. '23). (Plate V, figs. h and h') Wings curled upwards at tips, silvery or waxy in texture. Posterior scutellar bristles erect and sharply crossed. Body-color dark, especially on scutellum and the trident pattern.

Curly, Cy. II, to left of b. WARD ('23). Wings strongly bent upward. Nearly always lethal when homozygous. Associated with a CIIL and a CIIR.

c u r v e d, c. II, 75.5. B. (B. & S. '14). (Plate I, fig. b). Wings concave below, thin-textured, divergent, uplifted.

c u t, ct. I, 20.0. B. (U). (Figs. 14, 51). Wings cut to point, marginal vein and edge of wing scalloped. Eyes smaller and kidney-like. Antennae flattened, aristae concave forward. Lost. Cut 2, 4, 5, 6, 7, 8, 11, 12, 13, 14, 15, 16. Slightly different cut allel. found by B. (U.). Cut^6 is the most useful and the slightest departure from wild-type. Cut 3, 9, 10. Slightly different cut allel. found by M. Cut^3 is the most extreme, and produces an abnormal sex-ratio by being much more inviable in ♀.

d a c h s, d. II, 31.0. M. & B. (B. & M. '19). (Plate II, fig. d). Tarsi 4-jointed instead of 5-jointed. Legs short. Venation abnormal. Viability erratic.

d a c h s-l e t h a l. II, 31 ±. B. (B. & M. '19). Probably a deficiency for dachs locus. Reduces crossing-over in its neighborhood.

d a c h s o i d, II. S. (B. & M. '19). Flies much foreshortened as though from pressure. Wings short and extended. Venation abnormal. Bristles on anterior part of costa stand out. Discarded.

d a c h s o u s, ds. II, 14 ±. B. (U). Legs and wings short. Crossveins closer together. Mohr reports locus is 3 ±.

d a r k, B. ('19*c*). Specific modifier of eosin, makes the eye-color slightly darker. Lost.

d a r k, III. CLAUSEN (B. & M. '23). A brownish red eye-color, overlapping wild-type.

D e f o r m e d, Df. III, 47.5. CATTELL (B. & M. '23). (Plate V, fig. f). Eye smaller and pushed back and up. Tuft of bristles at lower anterior margin. Overlaps wild-type. Hom. form nearly always lethal.

D e l t a, Δ. III, 66.2. B. (B. & M. '23). (Fig. 31). Long. veins all expand into deltas where they join margin. Bristles more numerous and slightly irregular. Lethal when hom.

D e l t a-b, Δ-b, III. B. (B. & M. '23). More extreme than Δ. Lost.

d e l t e x, dx. I, 17 ±. B. (U). Looks like Delta in III, but is recessive non-lethal in X.

d e p r e s s e d. I, 18 ±. B. (M. & B. '16). Wings turned down sharply. Lost.

D e t a c h e d, Dt. II, 10 ±. B. (U). II-long. fails to reach margin. Overlaps wild-type.

D i c h a e t e, D. III, 40.4. B. (B. & M. '23). (Fig. 20a). Lethal when hom. In het. form, wings spread and uplifted. Anterior dorsocentrals usually missing.

D i c h a e t e m i n u s-m o d i f i e r s. II. S. (B. & M. '19). Modifiers tending to reduce number of bristles of Dichaete flies. Discarded.

D i c h a e t e p l u s-m o d i f i e r s. III, to right of rough. S. (B. & M. '23). Dominant modifier that tended to restore dorsocentrals normally removed by Dichaete. Discarded.

d i l u t e. III. B. (B. & M. '23). Very slight dilution of the red eye, somewhat more pronounced dilution of vermilion. Overlaps normal. Lost.

d i l u t e-o c e l l i. II. B. (U). Ocellar color pale. Class. difficult.

d i v e r g e n t, dv. III, 20.0. B. (B. & M. '23). (Fig. 25a). Wings spread, wings smaller and with slightly abnormal venation. ♀ rather sterile.

d o u b l e. I, 0.+. B. (U). Inner of the vertical bristles on top of head doubled. Wings very small. Viability slightly below standard. Lost.

d u m p y, T^{dy}. II, 13.0. M. (U). Recessive allel. of Truncate. Wings obliquely truncated at two thirds normal length. Thorax shows whorls and pits like vortex. Most useful T allel.

d u s k y, dy. I, 36.2. B. (U). (Fig. 14, cut [3]). A wing about 80 % the normal size. The texture darker than normal.

d u s k y², dy^2. I, 36.2. B. (U). A dusky allelomorph slightly less extreme.

d u s k y³, dy^3. I, 36.2. B. (U). A dusky allelomorph less extreme than dy and dy^2. Class. in females not entirely reliable. Discarded.

d w a r f, dw. III, 50 ±. B. (B. & M. '23). (Fig. 25b). Body-size about

75 % that of normal, overlaps normal. Associated with CIII. ♀-sterile.

d w a r f - b, dw-b. III, 12 ±. B. (B. & M. '23). Flies even smaller than dwarf, except in early counts. Discarded.

d w a r p, dwp. I, 0. 0. B. (U). Recessive dwarf with warped wings. Nearly always lethal. Special relation to speck.

e b o n y, e. III, 70.7. WALLACE (B. & M. '23). The darkest of the mutant body colors. Associated with CIII. Viability poor.

e b o n y ^{3}, e^3. III, 70.7. S. (B. & M.' 23). An e allel. intermediate between e and e^s. Lost.

e b o n y ^{4}, e^4. III, 70.7. S. (B. & M. '23). Very dark thorax color, abdomen rather lighter than in e. Viability rather poor.

e b o n y ^{5}, e^5. III, 70.7. B. (U). Intermediate between e^s and wild-type. Discarded.

e c h i n u s, ec. I, 5.5. B. (U). Eyes very large and rough.

é c r u, w^{ec}. I, 1.5. MULLER ('20*b*) Palest allelomorph of white, a faint yellow.

e o s i n, w^e. I, 1.5. M. (M. & B.'16). Eye-color a pinkish yellow in ♂, a yellowish pink in ♀. Allel. of white.

e x p a n d e d, ex. II, 6 ±. B. (U). Wings very broad, extended and curved downward. Viability poor.

E x t e n d e d, D^E. III, 40.4. S. (B. & M. '23). (Fig. 20*b*). Wings usually extended. Dorsocentrals normal. Lethal when hom. Allel. of D. Discarded.

e x t e n d e d-b. II, B. (U). Spread wings. Recessive.

E x t r a-s c u t e l l a r s-III. III, to right of pink. PAYNE (B. & M. '23). Polygenic character, increased number of scutellar bristles.

e y e l e s s, ey. IV, 0.9. HOGE ('15*b*). Eyes small or absent. Overlaps wild-type.

e y e l e s s^2, ey^2. IV, 0.9. NONIDEZ (U). (Plate VI, fig. d, d', d''). Eyes small or absent.

f a c e t, fa. I, 3.0. B. (M. & B. '16). Ommatidia irregular, more extreme in ♂, class. in ♀ unreliable.

f a t. II, 12 ±. MOHR ('23*c*). Body and wings short. Crossveins closer together.

f l i p p e r, fp. II, 30.0. MOHR ('19). Wings as in pupa, fly small, entirely sterile. Discarded.

f o r k e d, f. I, 56.5. B. (M. & B. '16). (Plate I, fig. c). All bristles gnarled, bifurcated or twisted, somewhat short.

forked^2, f^2. I, 56.5. MOHR ('22*a*). Posterior scutellars and few or many other bristles slightly twisted or forked. Class. unreliable.

forked^3, f^3. I, 56.5. B. (U). All bristles somewhat forked and twisted. More extreme than f^2. Class. difficult.

forked^4, f^4. I, 56.5. MOHR ('22*a*). All bristles more gnarled, forked and curled than in f. Also hairs curled distinctly as in sn.

(f-B)-deficiency. I, 56.5—57.0 inclusive. B. ('17*c*). Lost.

fringed, fr. II, 83.5. B. (B. & M. '19). Wings extended, marginal hairs very irregular. Viability rather poor.

furrowed, fw. I, 38.±. DUNCAN (M. & B. '16). Eyes much furrowed, bristles somewhat crooked especially on scutellum. Class. difficult.

fused, fu. I, 59.0. B. (M. & B. '16). (Fig. 23). III and IV long. veins fused together for part of their length. Ocelli reduced or absent. Bristles around ocelli reduced or absent. ♀ sterile.

fused^2, fu^2. I, 59.0. S. (U). Less extreme allel. of fu.

gap, gp. II, 74 ±. B. (B. & M. '19). (Plate II, fig. g). Section missing from IV-long. Overlaps wild-type except when intensifier is present.

gap-intensifier, gp-i. II, 30 ±. B. (U). The combination of gap and gap-intensifier is easily classifiable. This combination was formerly called "dash".

garnet, g. I, 44.4. B. ('16). A dark purplish-pink eye-color.

garnet^2, g^2. I, 44.4. B. (U) (PAYNE '21). An allelomorph of garnet that is a lighter and yellower pink.

garnet^3. g^3. I, 44.4. B. (U). Looks like g, but gives lighter orange combination with v.

Giant-II, GtII. II. B. (B. & M. '23). Lethal when hom. In het. form does nothing by itself, but in presence of hom. gtIII gives giant flies. ♂ sterile.

giant-III, III, 64.0. B. (B. & M. '23). A recessive gene that produces no effect by itself, but with GtII het. gives flies about 25 % heavier than wild-type. ♂-sterile.

glass, gl. III, 63.1. MULLER (B. & M. '23). (Plate IV, figs. a, b, c). Ommatidia of eye run together into smooth sheet as in lz^2. Eye small, slight yellowish pink in ♀, blotched vermilion in ♂.

Green, I. B. (M. & B. '16). Greenish dark body-color. Class. unreliable. Lost.

Gull, G. II, 12 ±. MOHR ('23*c*). Wings large, slightly curved, crossveins closer together. Probably a deficiency for fat.

Hairless, H. III, 69.5. B. (B. & M. '23). (Fig. 59). Postverticals, anterior dorsocentrals and abdominal bristles missing, leaving basal ring prominent. Long. IV and V often fail to reach margin. Lethal when hom.

hairy, h. III, 26.5. MOHR ('22a). (Plate II, figs. f and f'). Extra hairs along veins of wings, on scutellum, on head, and on mesopleura.

Hairy-wing, Hw. I, 0.0. ±. B. (U). Dominant. Extra hairs on wings, and other parts. In hom. form much more extreme, and more extreme in male.

Hairy-wing suppressor. III, 54.8. B. (U). Wings blunt, body squat. ♀-sterile. Recessive suppressor for the sex-linked dominant. Hairy-wing, even when Hw is hom.

humpy, hy. II, 90.0. B. (U). Wings short and truncated, thorax very humpy, with depressions and rosettes, as in vortex. ♀ rather sterile. Viability below par.

inflated, if, I, 55 ±. WEINSTEIN ('18). Wings blown up into balloons. Viability and fertility poor. Discarded.

Infra-bar, B^i. I, 56.5. S. (This paper, p. 30). Less extreme Bar. Eyes rough-surfaced.

ivory, w^i. I, 1.5. S. (MULLER '20*b*). A light pinkish-yellow eye-color somewhat paler than eosin, males slightly lighter. Allel. of white.

Jammed, J. II, 41.0. B. (U). Wings compressed into narrow strings, with venation partly eliminated. Lethal when hom.

jaunty, j. II, 46.7. B. (B. & M. '19). Wings turn up at ends. Occasionally hard to classify.

kidney, k. III, 64. B. (B. & M. '23). (Plate IV, fig. d). Nick in eye on anterior or lower margin, may extend to make eye very small. Overlaps wild-type.

lance. II. PAYNE ('20). Pointed wing. Overlaps wild-type.

lance-b. II, 107 ±. B. (U). Narrow pointed wing, overlaps wild-type.

lemon. I, 18 ±. WALLACE (M. & B. '16). A bright pale-yellow body-color. Viability poor. Lost.

Lethal-1, l1. I, 1.1. RAWLS (M. & B. '16). The first lethal found in Drosophila. Discarded

lethal-1a, l1a. I, 3.3. RAWLS (M. & B. '16). The first double lethal; it arose in a stock of l1, in the other X-chromosome. Lost.

lethal-1b, l1b. I, 1.5—. M. (M. & B. '16). Lost.

l e t h a l-2, 12, I, 16 ±. M. (M. & B. '16). Lethal that is allel. of club, or deficiency for cl. Reduces crossing-over around cl by about 5 units. Still being studied.

l e t h a l-3, 13, I, 26.5. M. (M. & B. '16). Discarded.

l e t h a l-3a, 13a. I, 19.5. B. (M. & B. '16). Second case of double lethal with nearly complete elimination of males. Lost.

l e t h a l-4, 14. I, 49 ±. B. (B. '16). Lost.

l e t h a l-5, 15. I, 41.7. B. (U). Lost.

l e t h a l-6, 16. I, —.04. B. (B. '16). Apparently to left of yellow by .04 units. Discarded.

l e t h a l-7, 17. I, 0.1. B. (B. '16; STARK '18). A gene that causes the development of one or more black bodies in that half of the male larvae which receives the gene. The death of these larvae may be brought about through the action of these growths.

l e t h a l-8, 18. I, 59.0 B. (U). A deficiency for fu. Decreased f-fu crossing-over from 2.4 to 0.6. Lost.

l e t h a l-9, 19. I, 39.5. B. (U). Discarded.

l e t h a l-10, 110. I, 0.+. B ('17*a*). A lethal that allows a rare survival as a fertile dwarf ♂. Lost.

l e t h a l-11, 111. I, 54.4. B. (U). Nearly complete lethal. Discarded.

l e t h a l-12, 112. I, 0±. B. (U). Discarded.

l e t h a l-13, 113. I, 22 ±. B. (U). Discarded.

l e t h a l-13a, 113a. I, 21 ±. B. (U). A double lethal that with 113 eliminates nearly all males. Lost.

l e t h a l-14, 114. I. B. (U). Discarded.

l e t h a l-15, 115. I, 13 ±. B. (U). Cuts down crossing-over to left of it. Discarded.

l e t h a l-16, 116. I, 37 ±. B. (U). Discarded.

l e t h a l-17, 117. I, to right of f. B. (U). Semi-lethal. Arose in XXY ♀ as mutation in one X but not in other. Lost.

l e t h a l-18, 118. I, 67±. S. (U). Discarded.

l e t h a l-19, 119. I, 26 9. MOHR ('23*l*). Discarded.

l e t h a l-20, 120. I, 44 ±. B. (U). Rarely hatches as a dilapidated dwarf. Discarded.

l e t h a l-21, 121. I, 50 ±. MOHR ('23*l*). Discarded.

l e r h a l-22, 122. I, 50 ±. B. (U). Discarded.

l e t h a l-23, 123. I, 40 ±. B. (U). Discarded.

l e t h a l-Ig. I, 63.1. BONNIER ('23*b*). Kills hom. ♀. Not lethal in ♂.

l e t h a l-Il. I, 7.5. BONNIER ('22). Arose in XXY ♀ in studies on high non-disjunction.

l e t h a l-Ir. I, 16.4. BONNIER ('22). Arose in XXY ♀ in studies on high non-disjunction.

l e t h a l-IR. I, 15.7. BONNIER ('22).

l e t h a l-br^a. I, 0.6. MULLER ('23).

l e t h a l-br^b. I, 0.6. MULLER ('23).

l e t h a l-e r e c t. I, 38 ±. THOMPSON ('21). In ♂ makes wings erect. In het. ♀ recessive; in hom. ♀ is lethal.

l e t h a l s i n X. MULLER and ALTENBURG ('19). About 50 sex-linked lethals, loci not reported. Discarded.

l e t h a l-sa. I, 23.7. STARK (M. & B.' 16). Lost.

l e t h a l-sb. I, 16.7. STARK (M. & B. '16). Lost.

l e t h a l-sc. I, 66.2. STARK (M. & B. '16). Discarded.

l e t h a l-sd. I. STARK (M. & B. '16). Lost.

l e t h a l-IIa, lIIa. II, 102 ±. B. (B. & M. '19). Probably deficiency. Decreases crossing-over in rt. limb of II.

l e t h a l-IIb, lIIb, II, 42 ±. B. (U). Discarded.

l e t h a l-e. II, 50 ± MULLER ('23), occurred 4 times. Three others (la, lb, lc) occurred twice.

l e t h a l o p p o s i t e S t a r. II, 15 ±. B. (U). Discarded.

l e t h a l-s q. II, 13 ±. B. (U). Probably allel. of Tr. Lost.

l e t h a l-Tr. Tr. II, 15 ±. M. (B. & M. '19). A semi-balancer for Truncate. Lost.

l e t h a l w i t h vg. II. L. J. COLE (U). Discarded.

l e t h a l s in II. MULLER ('23). About 30, loci unreported.

l e t h a l-IIIa, lIIIa. III, 83 ±. MULLER (B. & M. '23). Associated with CIII, used in balancing stocks.

l e t h a l-IIIb, lIIIb. III, p. ± 7. M. (B. & M. '23). Discarded.

l e t h a l-IIIc, lIIIc. III, 50 ±. Dexter (B. & M. '23). Lost.

l e t h a l-IIId, lIIId. III, about 33 units to rt. or left of p. Liff. (B. & M. '23). Lost.

l e t h a l-IIIe, lIIIe, III, 40.7. S. (B. & M. '23). Discarded.

l e t h a l-IIIf, lIIIf. III, 39.1. S. (B. & M. '23). Discarded.

l e t h a l-IIIg, lIIIg. III, 14 ± or 38 ±. B. (B. & M. '23). Discarded.

l e t h a l-IIIh, lIIIh. III, 22 ±. B. (B. & M. '23). Lost.

l e t h a l-IIIi, lIIIi. III, 63 ±. B. (B. & M. '23). Discarded.

l e t h a l-IIIj, lIIIj. III, 75 ±. B. (U). Lost.

lethal-head defect. III. B. (U). Recessive that killed most flies in pupa stage with black tumor-like growth in head. Lost.

lethal opposite Dichaete, III. MOHR ('23*b*).

lethal with CIIIL. L. V. MORGAN (U).

lethal with CIIIPR. PAYNE (In press).

lethal with CIIIPL. PAYNE (In press).

limited. II, 107 ±. B. (B. & M. '19). Abdominal bands somewhat smaller in size and irregular in outline. Probably only another effect of the morula gene.

Lobe, L. II, 72.0. B. (MOHR '23*d*). (Plate III, fig. e). In het. form, eye small or notched anteriorly. Overlaps wild-type. In hom. form eye much smaller. Best used as a recessive.

Lobe^2, L^2. II, 72.0. MOHR ('23*d*). A more extreme allelomorph than Lobe, class. reliable. Hom. form varies from het. form to eyeless. Hom. viable.

Lobe^3, L^3. II, 72.0. B. (U) More extreme than L, less than L^2. Not satisfactory as a recessive or as a dom. Discarded.

lozenge, lz. I, 27.7. B. (WEINSTEIN '18). A narrow eye, with ommatidia largely run together.

lozenge^2, lz^2. I, 27.7. B. (U). A much more extreme allel. of lozenge, in that eye is smaller, translucent brownish-red in color. All ommatidia run together into smooth hairless sheet. lz^2 vermilion is white in eye-color.

lozenge^3, lz^3. I, 27.7. B. (U). Lozenge allelomorph that looks much like lz^2 but does not give white eye-color with vermilion.

lozenge^4, lz^4. I, 27.7. B. (U). A less extreme lz allelomorph. Eyes somewhat smaller, uneven surface with patches of run-together ommatidia.

lozenge^5, lz^5. I, 27.7. B. (U). A very extreme allelomorph that arose from lz. Lost.

mahogany, my. III, 49.5. B. (B. & M. '23). Pinkish brown eye-color, overlaps wild-type. Discarded.

maroon, ma. III, 49.7. B. (B. & M. '23). Reddish pink eye-color. Tends to overlap wild-type in late hatches of cultures. Intensified by vermilion.

miniature, m. I, 36.1. M. (M. & B. '16). A wing little over half the normal size, very dusky in color.

Minute, M. III, 101.0. B. (B. & M. '23) (Plate V, fig. g). Bristles small. Hatches about 3 days later than wild-type. Lethal when hom.

M i n u t e-a, Ma. II, 107 ± S. (U). Very late hatching Minute, lost through sterility.

M i n u t e-b, Mb. II, 87 ±. B. (U). Formerly MII. A Minute with very small bristles and rather pronounced abnormality of abdomen, Lethal when hom. Lost.

M i n u t e-dII, MdII. II, 73 ±. B. (B. & M. '23). Formerly M5II. Lethal when hom. In het. produces no effect except in conjunction with MdIII, the double het. having minute bristles. Class. unreliable.

M i n u t e -dIII, MdIII. III, 95 ±. B. (B. & M. '23). A Minute with little effect unless intensified by MdII. Lethal when hom.

M i n u t e-e, Me. II, 46 ±. B. (U). Formerly M6. A slight Minute, but accurately classifiable. Lethal when hom.

M i n u t e-f, Mf. III, 105 ±. B. (B. & M. '23). Discarded.

M i n u t e-g, Mg. III, 196.2. B. (B. & M. '23). For successful class. requires intensifier near pink. Lethal when hom.

M i n u t e-g i n t e n s i f i e r, I-Mg. III, near pink. B. (U). Dominant intensifier of Mg. Does nothing by itself.

M i n u t e-h, Mh. III, 40.1. Mrs. S. (B. & M. '23). Excellent viability and separability. Lethal when hom.

M i n u t e-i, Mi. III, 30 ±. B. (U). Lethal when hom. Lost.

M i n u t e-j, Mj. III, 90 ±. B. (U). Lethal when hom.

M i n u t e-k, Mk. I, 36.3 ±. B. (U). A Minute that is dominant in ♀, lethal in ♂.

M i n u t e-l, Ml. II, 99.5—101.5 inclusive. B. (U). A deficiency for arc and plexus.

M i n u t e-m, Mm. II, 45 ±. B. (U). Lethal when hom.

M i n u t e-n, Mn. I, 62 ±. B. (U). A minute that is dom. in ♀, lethal in ♂.

M i n u t e-o, Mo. I, 56 ±. B. (U). A Minute that is dom. in ♀ and lethal in ♂.

M i n u t e-p, Mp. II, 53 ±. B. (U).

M i n u t e-q, Mq. III, 43 ±. B. (U). Discarded.

m o r u l a, mr. II, 107.0 ±. B. (B. & M. '19). Eyes very rough, bristles reduced in size. ♀ totally sterile.

n a r r o w, nw. II. B. (U). A narrow wing slightly folded longitudinally.

n i c k, vg^{n}. II, 67.0. B. (B. & M. '19). Allelomorph of vg that produces no effect by itself, vg-vg^n compound usually has a nick at ends of wings. Discarded.

N o t c h, N. I, 3 ±. DEXTER ('14). (Fig. 18). In het. form wings have terminal notch, usually at tips of III and IV long. Long. III and V are thickened throughout. Hairs and bristles more numerous and irregular. Notches (1, 3, 6, 7, 10, 11, 13, 15, 16, 17, 18, 19, 20, 21, 22, 23), slightly different Notches found by B. (U.) Notch-2, MULLER. Notches (4,9) M. Notches (8, 12), MOHR. Notch-14, S. Notches 8, 9, 18, and an unnumbered one found by L. V. MORGAN are known to be deficiency for white locus. Notches (8, 18, 19) still on hand.

N o t c h -b, II. MANN ('23*a*). Formerly "Notch 2". Slight notch in one or rarely both wings. Character shows in less than a third of het. individuals. Discarded.

N o-w i n g s. II. M. (U). Wings and balancers reduced as in apterous. Scutellar bristles as in vestigial. Female rarely fertile. (Fig. 26).

o b l i q u e, T°. II. B. (U). Wing obliquely truncated near the tip.

O f f, O. II, 62 ±, or 82 ±. B. (U). In het. form bristles gone, leaving rings, especially from sides of abdomen. In hom., more bristles gone and eyes rough. Can be used as dominant (class. difficult) or recessive (class. easy).

o l i v e-II, olII. II. M. (B. & M. '19). Dark trident pattern. Discarded.

o l i v e-III, III. M. (B. & M. '23). Discarded.

o r a n g e, o. III, to right of pink. B. (U). Eye-color transparent orange. Associated with CIIIOL, CIIIOR.

o v a l e s s. II. B. (U). ♀ entirely without ovaries, ducts and genitalia normal. ♂ normal. Lost.

p a d s, ps. II, far from sp. B. (U). (Plate II, fig. b). Wings often as in pupa, usually more expanded.

p a d s-b, II. DUNCAN (U). Wings as in pupa.

P a l e-II, PII, 99.7—107.5 inclusive. B. (B. & M. '23). A deficiency for the right end of chrom. II.

P a l e-III, III, 80 ±. B. (B. & M. '23). A translocated section of chromosome II covering plexus to the end. Eliminates most of the crossing-over in III from stripe to the right end. Lethal when hom.

p a r t e d, pt. II, 42 ±. B. (U). Hairs on thorax parted, with clear strip down mid-thorax. Wings spread. ♂-sterile, penis rotated through 180°.

p a t c h e d. II. B. (B. & M. '19). Abdominal bands triangular and fitted together obliquely. Discarded.

p e a c h, p^p. III, 48.0. B. (B. & M. '23). yellowish-pink transparent eye-color. Easier to work with than pink. Allel. of pink.

p i n k, p. III, 48.0. M. (B. & M. '23) Eye-color pink.

p i n k^3, p^3. III, 48.0. B. (B. & M. '23). Light pink eye-color. Lost.

p i n k^4, p^4. III, 48.0. RICHARDS (B. & M. '23). A lighter allel. of pink. Lost.

p i n k^5, p^5. III, 48.0. B. (U). Darker than pink. Discarded.

p i n k i s h. II, B. (B. & M. '19). Specific dilutor of eosin. Partially sex-limited. Discarded.

p i n k o i d. II. LANCEFIELD ('18*d*). Bright pink eye-color. Lost.

p i n k-w i n g, pw. II, 64 ±. B. & S. (U). (Plate III, fig. c). Eye-color pink, wings smaller, discolored, extended slightly.

P l e x a t e, Px. II, 107 ±. B. (U). Slight plexus of veins. Lethal when hom. Deficiency for sp.

p l e x u s, px. II, 100. 5. B. (B. & M. '19). (Plate II, fig. f). Many branched and extra veins.

P o i n t e d-w i n g, Pw. III, 94.1. B. (B. & M. '23). (Fig. 28). Wings pointed. Long. III and IV veins approach at margin. Class. difficult. Lethal when hom.

p o rt. III, M. (B. & M. '23). Pink eye-color. Lost.

p o r t-b. III. B. (B. & M. '23). Pinkish eye-color, overlaps wild-type. Fails to modify color of eosin. Lost.

p r o n g, pg. II, 40 ±. MOHR ('23*d*). Extra crossvein, usually incomplete, distal to the anterior crossvein. Class. not reliable.

p r u n e, pn. I, 1.0. B. (U). Transparent brownish-red eye-color.

p u r p l e, pr. II, 54.5. B. (B. & M. '19) purplish-pink eye-color.

p u r p l e^2, pr^2. II, 54.5. L. V. MORGAN (U). Slightly more reddish-purple allel. Discarded.

p u r p l e o i d, pd. II, 106 ±. B. (U) dark purplish-pink eye-color.

r e d u c e d, rd. II, 52 ±. B. (U). Bristles smaller, but irregularly so. ♀-sterile. In ♂ bristles very small.

r e d u p l i c a t e d. I, 34.7. HOGE (M. & B. '16). Legs reduplicated either throughout or with respect to terminal segments. At room temperatures few genotypically reduplicated flies showed the character; at 10° C. nearly all did. Most extreme individual showed parts of 14 legs. Discarded.

r o l l e d. II. B. (U). Edges of wing rolled, concave below.

r o o f. II, 79 ±. B. (U). Wings slope sharply at sides. Discarded.

R o o f-c, Rfc. III, to left of se. B. (B. & M. '23). Wings slope to sides. Lost.

r o s e, rs. III, 35.0. B. (U). Purplish pink eye-color. ♂-sterile. Rather inviable.

r o t a t e d-a b d o m e n III. B. (B. & M. '23). (Fig. 21). Tip of abdomen rotated to left through 60 to 90 degrees. Both sexes unable to mate.

r o u g h, ro. III, 91.1. Muller (B. & M. '23). Eye with hairs and ommatidia disarranged.

r o u g h^2, ro^2. III, 91.1. Muller (B. & M. '23). More extreme ro allel. Discarded.

r o u g h i s h. I, 12 ±. B. (U). Eye rough like Star. Viability rather poor. Lost.

r o u g h o i d, ru. III, 0.0. S. (B. & M. '23). (Plate V, Figs. a, b, c, d). Eye slightly smaller, rough by disarrangement of facets. Slight smudge of blackened facets in lower part of eye.

r o u g h o i d^2, ru^2. III, 0.0. B. (U). Slight allel. of ru. No smudge present.

r u b y, rb. I, 7.5. B. (U). Transparent pink eye.

r u b y^2, rb^2. I, 7.5. B. (U). Ruby allelomorph; darker and more purplish-pink than ruby.

r u d i m e n t a r y, r. I, 54.5. M. (M. & B. '16). (Fig. 24). Wing obliquely truncated at $\frac{2}{3}$ the normal length, slightly arched, III-long. has break. Marginal bristles on wing stand out and are irregular. Rudimentary (2, 5, 6, 7, 10, 13), B. (3, 8, 11), S. (4) Wallace. Slightly different allelomorphs. Discarded or lost.

r u d i m e n t a r y^9, r^9. I, 54.5. B. (U). A rudimentary allel. much less extreme than r. ♀ fairly fertile with r ♂.

r u d i m e n t a r y^{12}, r^{12}, I, 54.5. Wallace (U). Even less extreme than r^9; r^{12} ♀ still more fertile. Class. not dependable. in ♂.

r u d i m e n t a r y^{14}, r^{14}. I, 54.5. B. (U). An extreme r allel. Wings usually less than half normal length, much blistered, and with venation abnormal.

s a b l e, s. I, 43.0. B. (M. & B. '16). Body-color darker, trident-pattern very dark and sharp. Viability somewhat poor.

s a b l e^2, s^2. I, 43.0. B. (U). A less extreme allelomorph of sable, viability excellent.

s a b l e-d u p l i c a t i o n, I, 0.0. B. ('19*a*). Recessive suppressor for sable; allelomorph also to (v,s)-dup.

s a f r a n i n, sf. II, 60 ±. B. (U). Dark pinkish-brown eye-color.

s a f r a n i n-b.. III. M. (B. & M. '23). Brownish-pink eye-color that overlapped wild-type. Lost.

s c a r l e t, st. III, 44.0. HOGE (B. & M. '23). Vermilion eye-color, opaque. With age it becomes darker than vermilion or cinnabar.

s c r a g g l y, rd^s. II, 52 ±. B. (U). Allel. of reduced. ♀-fertile. Sex-lim.

s c u t e, sc. I, 0.+ B. (PAYNE '20). (Fig. 14, cut 6. Plate I, fig. c). Bristles missing, most frequently the posterior scutellars, the ocellars, and the postverticals.

s c u t e-d u p l i c a t i o n. I, 70 ±. L.V. MORGAN (U). Suppresses sc, y, Hw, and br. Reduces crossing-over from cut progressively more throughout right end of X.

s c u t e-i n h i b i t o r. III. PAYNE (B. & M. '23). A dominant in the right limb of III, that tends to restore the bristles normally removed by scute.

s e m i-f o r k e d. III. LANCEFIELD (B. & M. '23). A recessive that made heterozygous forked a semi-dominant. Discarded.

s e p i a, se. III, 26.0. WALLACE (B. & M. '23). Very dark brownish-pink eye-color, deepening with age to black.

S h a v e n, sv. IV, 0.5 ±. B. (U). Plate VI, fig. c). Bristles absent from thorax and abdomen. More extreme in ♀.

s h i f t e d. I, 17.8. B. (M. & B. '16). III and IV long. veins shifted closer together. Anterior c-v absent. Lost.

s h o r t, br^s. I, 0.6. B. (B. '16). More extreme than broad, wings often arched or wrinkled. Lost. Allel. of broad.

s h o r t-w i n g. III, 60 ±. B. (U). Wings like dy in shape but not as dark.

s h r u n k e n, III, near pink. B. (B. & M. '23). Small flies, much dilapidated. Lost.

s i l v e r. I, 0.1 ±. B. (U). pale yellow body-color, legs silvery, viability fair only, possibly an allel. of chlorotic.

s i n g e d, sn. I, 21.0 MOHR ('22*a*). All bristles and hairs are twisted and gnarled. The ♀ is entirely sterile, the eggs being small and abnormally shaped.

s i n g e d^2, sn^2. I, 21.0 B. (U). All bristles slightly twisted and gnarled, hairs normal. ♀-fertile. Class. somewhat unreliable. The sn^2 of MOHR ('22) is not this sn^2 but is a reappearance of sn.

s i n g e d^3, sn^3. I, 21.0. (MOHR '23*a*). All bristles and hairs twisted and gnarled to the same degree as in singed. But ♀ sn^3 is fully fertile.

S k i-II, Si. II, 35.0. CLAUSEN (B. & M. '23). (Fig. 16). By itself produces no effect. In presence of hom. ski-III, het. Si gives sharply upturned wings, and hom. Si gives character slightly more extreme and with an infold in the marginal vein near the break in the costa.

s k i-III, si. III, 46.5. CLAUSEN (B. & M. '23). Wings upturned at end, more than in jaunty, less than in curled and Curly. Character dependent upon hom. si, het. Si. When both Si and si are hom. the curling is somewhat more pronounced and there is an infolding of the costal vein opposite the anterior crossvein.

s m a l l-e y e, sy. I, 58.5 B. (U). Eye small and rounded, situated high on the head but not protuberent. Class. accurate but slow.

s m a l l-w i n g, sl. I, 54.2. B. (U). Wing about 80 % normal length, anterior margin straighter than normal, crossveins somewhat closer.

s m u d g e, sm. III, 15 ±. B. (B. & M. '23). Black smudge on upper part of eye consists of ommatidia with black cone-shaped eruption. Hairs on eye also darker. Class. poor. Discarded.

S n u b, T^s. II, 13 ±. B. (B. & M. '19). Truncate allel. Discarded.

s o o t y, e^s. III, 70.7. S. (B. & M. '23). An allel. of e not so dark. In some experiments separation is not reliable.

s p e c k, sp. II, 107.0. B. (B. & M. '19) (Plate I, fig. d). Black spot in axil of wing.

s p i n e l e s s, ss. III, 58.5. B. (B. & M. '23). (Plate I, fig. a). All bristles very short.

s p i n e l e s s-S u p p r e s s o r. III, 64 ±. B. (U). Lethal when hom. In het. form turns hom. ss into wild-type except for posterior scutellars, which remain small.

s p o o n. I, 0.2. ±. B. ('16). Thin-textured curved wing with extra crossvein, a semi-lethal that allows only about 10 % of the spoon flies to hatch. Fertile. Lost.

s p o t, y^s. I, 0.0. E. CATTELL (M. & B. '16) y allel., differs from y in having very light yellow spot on dorsal tip of abdomen, and light streak down thorax. Discarded.

s p r e a d, sd. III, 65 ±. DEXTER (B. & M. '23). Wings spread, associated with CIII. Lost.

S q u a t, Sq. II, 38 ±. B. (B. & M. '19). Wing and body short and broad. Class. difficult. Discarded.

s q u a r e. II, 13 ±. B. (U). Probably non-lethal allel. of T. Lost.

S t a r, S. II, 2.0. B. (B. & M. '19). (Plate III, figs. c, f, and g.). Eyes small and rough, with hairs and ommatidia irregular. Lethal when hom.

S t a r-i n t e n s i f i e r, S-i. III, 10 ±. B. (B. & M. '23). Discarded.

s t r a p, vg^s. II, 67.0. M. (B. & M. '19). (Fig. 15). An allel. of vg., varying from vg-like wings to long slender mid-regions of wings. Discarded.

s t r a w. II, near black. B. (U). Recessive, slightly yellow body-color. Class. difficult.

S t r e a k, Sk. II, 16.0. B. (B. & M. '19). Dark streak down middle of thorax. Easiest classified in black flies. Lethal when hom.

s t r i p e, sr. III, 62.0. B. (B. & M. '23). Dark stripe down middle of thorax. Shows best in dark body-color flies.

S t u b b l e, Sb. III, 58.2. B. (U). All bristles cut off abruptly at half the normal length. Lethal when hom.

s u p e r w i t h. III. M. (B. & M. '23). Lost.

t a n, t. I, 27.5. B. (U). Body color yellower than wild, especially the abdomen. Easiest classified by antennae, which are light yellow, sex-limited: class. in males difficult, in females so difficult as to be unreliable. Over-etherization, or prolonged etherization, turns the antennae grey or black, so tan should be classified first, rapidly, and under light etherization. Indifferent to light (McEWEN '17).

t a n², t^2. I, 27.5. B. (U). Less extreme tan than the original allelomorph. Class. unreliable. Not indifferent to light.

t a n- s u p p r e s s o r. III, 50 ±. B. (U). In hom. form suppresses hom. tan. Probably increases crossing-over in middle of III.

t e l e g r a p h, tg. II, 0.0. B. (U). Break in IInd long. vein.

t e l e s c o p e, ts. II, 68 ±. B. (B. & M). Abdominal segments somewhat drawn out. Wings divergent and drooping. Class. difficult.

T h o r a x a t e, T^x. II, 13 ±. B. (U). Allel. of T., by itself is lethal when hom. but has no somatic effect in het. T^x-T^{dy} compound shows vortex character.

t h r e a d, th. III, 42.2. B. (U). Arista without lateral branches.

t i l t, tt. III, 40.2. B. (B. & M. '23). Wings tilted upward slightly at tips. Usually wing extended and uplifted. Usually break in III-long.

t i n g e d, w^t. II, 1.5. HYDE ('16). Very light eye-color, yellowish pink. Allel. of white.

tiny-bristle, tb. I, 36.0. B. (U). All bristles small. Viability rather poor. Lost.

translucent. II. B. (U). Slightly translucent pinkish-red eye-color.

trefoil, tf. II, 63 ±. M. (B. & M. '19). Dark pattern on thorax and head. Discarded.

Truncate, T. II, 13 ±. M. (B. & M. '19). Wings obliquely truncated near end. Overlaps wild-type. Lethal when hom.

Truncate-intensifier-III. MULLER (B. & M. '19). Lost.

tumor-b III. B. (U). A benign tumor.

Two-bristle, 2b. III, 58.3. B. (B. & M. '23). Anterior dorso-centrals and posterior verticals missing, no basal ring left. Lost.

Ultra-Bar, or Double-Bar, BB. I, 57.0. (ZELENY '20). See also this paper pages 29—32. (Fig. 13c). Very small Bar eye.

varnished, vr. III, 44 ±. MOHR (B. & M. '23). Eyes with facets run together. ♀-sterile. Lost.

vermilion, v. I, 33.0. M. (M. & N. '16). Vermilion eye-color, not translucent.

vermilion-deficiency. I, 33 ±. B. ('19*d*). Lost.

(v,s)-duplication I, 0.0 B. ('19*a*). In hom. form acts as suppressor for both vermilion and sable, but not for mutants located between.

vesiculated, vs. I, 15 ±. MOHR (U). Wing thin-textured, darker, crumpled, a vesicle in one or both wings, centering about long. IV beyond the posterior cv.

vestigial, vg. II, 67.0. M. (B. & M. '19). (Fig. 15). Wings and balancers reduced greatly. Posterior scutellar bristles erect.

vortex-II, voII. II, 13 ±. B. (B. & M. '19). (Fig. 17). Whorls of hairs and bristles, with crater. Probably allel. of T. By itself the voII does nothing. In flies hom. for voIII gives slight amount of vortex character. Hom. voII hom. voIII gives full vortex.

vortex-III, voIII. B. (B. & M. '23). Character vortex has whorls of hairs and bristles about pits or eruptions on thorax. Digenic character, dependent on hom. III-chrom. recessive and II-chrom. semi-dominant.

warped, wp. III, 46 ±. B. (B. & M. '23). Compound curvature to wings, which are smaller than normal, extended and uplifted and discolored.

white, w. I, 1.5. M. (M. & B. '16). Snowy white eye-color

white-head. III. M. (B. & M. '23). Silvery white patch around ocelli or along margin of eyes. Overlapped normal. Discarded.

white-ocelli, wo. III, 76.2. B. (B. & M. '23). Ocelli colorless.

whiting. III. B. ('19*c*). Specific modifier of eosin, giving a pure white eye. Lost.

With. III, 48 ±. M. (B. & M. '23). Semi-dominant dark trident pattern on thorax. Discarded.

yellow, y. I, 0.0. WALLACE (M. & B. '16) yellow body-color.

yellowish, sw^y II, near black. B. (U). Recessive yellowish body color. Allel. of straw.

LITERATURE

Adolph, E. F., '20. Egg-laying reactions in the pomace fly, Drosophila. Journ. Exper. Zool., 31: 327—341.

Aldrich, J. M., '05. A catalogue of North American Diptera (or two-winged flies). Smithson. Misc. Coll. 46, no. 1444. 680 pp. (Drosophilidae, pp. 639—644. Drosophila, pp. 641—644).

Altenburg, E., and H. J. Muller, '20. The genetic basis of truncate wing, an inconstant and modifiable character in Drosophila. Genetics, 5: 1—59.

Aubertot, A., '23. Sur la dissémination et le transport de Nématodes du genre Rhabditis par les Dipteres. C. R. Acad. Sci. 176: 1257—1260.

Austen, E. E., '05. A Dipterous enemy of English hot house grapes. Ent. Mo Mag., 276—278.

Babcock, E. B., and R. E. Clausen, '18. Genetics in relation to agriculture. McGraw-Hill Book Co. pp. 1—675.

Bacot, A. W., and A. Harden, '22. Vitamin requirements of Drosophila. I, Vitamins B and C. Biochem. Journ., 16: 148—152.

Baerg, W. J., '20. An unusual case of parasitism on Clastoptera obtusa. Ent. News, 31: 20—21.

Barrows, W. M., '07. The reactions of the pomace fly, Drosophila ampelophila, to odorous substances. Journ. Exp. Zool., 4: 515—537.

Baumberger, J. P., '14. Studies on longevity of insects. Ann. Ent. Soc. Amer., 7: 323—352.

Baumberger, J. P., '17*a*. The food of Drosophila melanogaster Meigen. Proc. Nat. Acad. Sci., 3: 122—126.

Baumberger, J. P., '17*b*. Solid media for rearing Drosophila. Amer. Nat., 51: 447—448.

Baumberger, J. P., '19. A nutritional study of insects with special reference to microorganisms and their substrata. Journ. Exp. Zool., 28: 1—81.

Baumberger, J. P., and R. W. Glaser, '17. The rearing of Drosophila ampelophila Loew on solid media. Science, n. s., 45: 21—22.

Bonnier, G., '22. Double sex-linked lethals in Drosophila melanogaster. Acta Zool., 3: 135—152.

Bonnier, G., '23*a*. Studies on high and low non-disjunction in Drosophila melanogaster. Hereditas, 4: 81—110.

Bonnier, G., '23*b*. Uber die Realisierung verschiedener Geschlechtsverhältnisse bei Drosophila melanogaster. Zeits. ind. Abst. Vererb. 30: 283—284.

Bonnier, G., '23*c*. On different sex-ratios in Drosophila melanogaster. Zeits. ind. Abst. Vererb. 31: 153—169.

Bonnier, G., '24. Contributions to the knowledge of intra- and interspecific relationships in Drosophila. Acta Zool., 5: 1—122.

Bowles, G. J., 1882. The pickled fruit fly — Drosophila ampelophila Loew. Canad. Entomol., 14: 101—104.

Bridges, C. B., '13. Non-disjunction of the sex-chromosomes of Drosophila. Journ. Exp. Zool., 15: 587—606.

Bridges, C. B., '14. Direct proof through non-disjunction that the sex-linked genes of Drosophila are borne by the X-chromosome. Science, n. s., 40: 107—109.

Bridges, C. B., '15. A linkage variation in Drosophila. Journ. Exp. Zool., 19: 1—21.

Bridges, C. B., '16. Non-disjunction as proof of the chromosome theory of heredity. Genetics, 1: 1—52, 107—163.

Bridges, C. B., '17*a*. The elimination of males in alternate generations of sex-controlled lines. Anat. Rec., 11: 504.

Bridges, C. B., '17*b*. An intrinsic difficulty for the variable force hypothesis of crossing over. Amer. Nat., 51: 370—373.

Bridges, C. B., '17*c*. Deficiency. Genetics, 2: 445—465.

Bridges, C. B., '18. Maroon — a recurrent mutation in Drosophila. Proc. Nat. Acad. Sci., 4: 316—318.

Bridges, C. B., '19*a*. Duplications. Anat. Record, 15: 357.

Bridges, C. B., '19*b*. The genetics of purple eye color in Drosophila melanogaster. Journ. Exp. Zool., 28: 265—305.

Bridges, C. B., '19*c*. Specific modifiers of eosin eye color in Drosophila melanogaster. Journ. Exp. Zool., 28: 337—384.

Bridges, C. B., '19*d*. Vermilion-deficiency. Journ. Gen. Physiol., 1: 645—656.

Bridges, C. B., '19*e*. The developmental stages at which mutations occur in the germ tract. Proc. Soc. Exp. Biol. Med., 17: 1—2.

Bridges, C. B., '20*a*. White-ocelli — an example of a "slight" mutant character with normal viability. Biol. Bull., 38: 231—236.

Bridges, C. B., '20*b*. The mutant crossveinless in Drosophila melanogaster. Proc. Nat. Acad. Sci., 6: 660—663.

Bridges, C. B., '21*a*. Gametic and observed ratios in Drosophila. Amer. Nat., 55: 52—61.

Bridges, C. B., '21*b*. Proof of non-disjunction for the fourth chromosome of Drosophila melanogaster. Science, n. s., 53: 308.

Bridges, C. B., '21*c*. Current maps of the location of the mutant genes of Drosophila melanogaster. Proc. Nat. Acad. Sci., 7: 127—132.

Bridges, C. B., '21*d*. Genetical and cytological proof of non-disjunction of the fourth chromosome of Drosophila melanogaster. Proc. Nat. Acad. Sci., 7: 186—192.

Bridges, C. B., '21*e*. Triploid intersexes in Drosophila melanogaster. Science, n. s., 54: 252—254.

Bridges, C. B., '22. The origin of variations in sexual and sex-limited characters. Am. Nat., 56: 51—63.

Bridges, C. B., '23. Aberrations in chromosomal materials. Eugenics, Genetics, and the Family, 1: 76—81.

Bridges, C. B., and O. L. Mohr, '19. The Inheritance of the mutant character "vortex". Genetics, 4: 283—306.

Bridges, C. B., and T. H. Morgan, '19. The second-chromosome group of mutant characters. Carnegie Inst. Wash. publ., 278: 123—304.

Bridges, C. B., and T. H. Morgan, '23. The third-chromosome group of mutant characters of Drosophila melanogaster. Carnegie Inst. Wash., publ., 327: 251 pp.

Bridges, C. B., and A. H. Sturtevant, '14. A new gene in the second chromosome of Drosophila and some considerations on differential viability. Biol. Bull., 26: 205—212.

Carpenter, F. W., '05. The reactions of the pomace fly (Drosophila ampelophila, Loew) to light, gravity, and mechanical stimulation. Amer. Nat., 39: 157—171.

Carpenter, F. W., '08. Some reactions of Drosophila, with special reference to convulsive reflexes. Journ. Comp. Neur. Psych., 18: 483—491.

Castle, W. E., '15. Mr. Muller on the constancy of Mendelian characters. Amer. Nat., 49: 37.

Castle, W. E., '19*a*. Is the arrangement of the genes in the chromosome linear? Proc. Nat. Acad. Sci., 5: 25—32.

Castle, W. E., '19*b*. The linkage system of eight sex-linked characters of Drosophila virilis (data of Metz). Proc. Nat. Acad. Sci., 5: 32—36.

Castle, W. E., '19*c*. Are genes linear or non-linear in arrangement? Proc. Nat. Acad. Sci., 5: 500—506.

Castle, W. E., '20. Model of the linkage system of eleven second chromosome genes of Drosophila: Proc. Nat. Acad. Sci., 6: 73—77.

Castle, W. E., F. W. Carpenter, A. H. Clark, S. O. Mast and W. M. Barrows, '06. The effects of inbreeding, cross breeding, and selection upon the fertility and variability of Drosophila. Proc. Amer. Acad. Arts and Sci., 41: 729—786.

Chambers, R., Jr., '14. Linkage of the factor for bifid wing. The bifid wing and other sex-linked factors in Drosophila. Biol. Bull., 27: 151.

Chatton, E., '12. Leptomonas roubaudi n. sp., parasite des tubes de Malpighi de Drosophila confusa Staeger. C. R. Soc. Biol., 73: 289—291.

Chatton, E., and E. Alilaire, '08. Coexistence d'un Leptomonas et d'un Trypanosoma chez un Muscide non vulnérant, Drosophila confusa. C. R. Soc. Biol., 64: 1004.

Chatton, E., and A. Krempf, '11. Sur le cycle évolutif et la position systématique des Protistes du genre Octosporea. Bull. Soc. Zool., 36: 172—179.

Chatton, E., and A. Léger, '11*a*. Eutrypanosomes. Leptomonas et Leptotrypanosomes chez Drosophila confusa (Muscide). C. R. Soc. Biol., 70: 34.

Chatton ,E., and A. Léger, '11*b*. Sur quelques Leptomonas de Muscides et leurs Leptotrypanosomes. C. R. Soc. Biol., 70: 120—122.

Chatton, E., and A. Léger, '11*c*. Documents en faveur de la pluralité des espèces chez les Leptomonas des Drosophiles. C. R. Soc. Biol., 71: 663—666.

CLAUSEN, R. E., '23. Inheritance in Drosophila hydei. I. Amer. Nat., 57: 52—58.

CLAUSEN, R. E., '24. The inheritance of cinnabar eye-color in Drosophila melanogaster, including data on the locus of jaunty. Journ. Exp. Zool., 38: 423—436.

CLAUSEN, R. E., and J. L. COLLINS, '22. The inheritance of ski wings in Drosophila melanogaster. Genetics, 7: 385—426.

COLE, W. H., '17. The reactions of Drosophila ampelophila Loew to gravity, centrifugation, and air currents. Journ. Anim. Behav., 7: 71—80.

COLE, W. H., '22. Note on the relation between the photic stimulus and the rate of locomotion in Drosophila. Science, n. s., 55: 678—679.

COMSTOCK, J. H., 1893. Report on miscellaneous insects. Report Comm. Agr., 1881—1882: 135—154.

CONGDON, E. D., '12. Effects of radium on living substance. The influence of radiations of radium upon the embryonic growth of the pomace fly Drosophila ampelophila. Bull. Mus. Comp. Zool., 53: 345—358.

CRAMPTON, G. C., '21. An exception to Dollo's law of the irreversibility of evolution. Science, n. s., 54: 91—92.

CROMWELL, H., '17. Further experiments with the mutant, scarlet, from Drosophila repleta. Proc. Ind. Acad. Sci., '17: 287—293.

CUÉNOT, L., and L. MERCIER, '23. Les muscles du vol chez les mutants alaires des Drosophiles. C. R. Acad. Sci. 176: 1112—1113.

DELCOURT, A., '09. Sur l'apparition brusque et l'hérédité d'une variation chez Drosophila confusa. C. R. Soc. Biol., 66: 709.

DELCOURT, A., '11. Sur un procédé permettant l'examen à un fort grossisement, a l'état vivant, de mouches de petite taille, notamment de Drosophiles. C. R. Soc. Biol., 70 : 97.

DELCOURT, A., and E. GUYÉNOT, '10. De la possibilité d'étudier certains Diptères en milieu défini (Drosophila). C. R. Acad. Sci. Paris, 5: 255—257.

DELCOURT, A., and E. GUYÉNOT, '11. Variation et milieu. Lignées de Drosophiles en milieu stérile et défini. C. R. IV Conf. Int. Genet.: 478—487.

DELCOURT, A., and E. GUYÉNOT, '11. Génétique et milieu. Nécessité de la détermination des conditions; sa possibilité chez les Drosophiles. Technique. Bull. Scient. France Belg., 45: 249—333.

DETLEFSEN, J. A., '20. Is crossing over a function of distance? Proc. Nat. Acad. Sci., 6: 663—670.

DETLEFSEN, J. A., and L. S. CLEMENTE, '23. Genetic variation in linkage values. Proc. Nat. Acad. Sci., 9: 149—156.

DETLEFSEN, J. A., and E. ROBERTS, '21. Studies on crossing-over. I. The effect of selection on crossover values. Journ. Exp. Zool., 32: 333—354.

DEXTER, J. S., '12. On coupling of certain sex-linked characters in Drosophila. Biol. Bull., 23: 183.

DEXTER, J. S., '14. The analysis of a case of continuous variation in Drosophila by a study of its linkage relations. Amer. Nat., 48: 712—758.

DUDA, O., '21. Kritische Bemerkungen zur Gattung Scaptomyza. Ver. schles. Insektenk., 13: 1—13.

DUDA, O., '22. Liodrosophila und Sphaerogastrella, zwei neue, zu den Drosophiliden und nicht zu den Camilliden gehörige Dipteren-Gattungen aus Südostasien. Arch. Naturgesch. 88A, Heft 4: 151—160.

DUFOUR, L., 1839. Les métamorphoses de plusieurs larves fongivores appartenant à les Diptères. Ann. Sci. Nat., ser. II, 12: 5—60. Pl. I—III.

DUNCAN, F. N., '15*a*. A note on the gonads of gynandromorphs of Drosophila ampelophila. Amer. Nat., 49: 455.

DUNCAN, F. N., '15*b*. An attempt to produce mutations through hybridization. Amer. Nat., 49: 575—582.

ELWYN, A., '17. Effect of humidity on pupal duration and on pupal mortality of Drosophila ampelophila Loew. Bull. Amer. Mus. Nat. Hist., 37: 347—353.

ESCHER—KÜNDIG, J., '03. (Note on Drosophila phalerata Meigen bred from larvae found in a human tumor). Mitt. Schweiz. ent. Ges., 10: 446—448.

FALLÉN, C. F., 1823. Diptera Sueciae. Geomyzides. vol. 2.

FERRY, R. M., R. C. LANCEFIELD, and C. W. METZ, '23. Additional mutant characters in Drosophila willistoni. Jour. Hered., 14: 372—384.

FISHER, R. A., '22. The systematic location of genes by means of crossover relations. Amer. Nat., 56: 406—411.

FREY, R., '18. Bananflugan (Drosophila ampelophila Loew) anträffand i Finland. Medd. Soc. Faun. Flor. Fenn. 44: 120—122.

FREY, R., '21. Studien über den Bau des Mundes der niederen Diptera Schizophora. Act. Soc. Faun. Flor. Fenn., 48, No. 23: 245 pp. X plates.

GOLDSCHMIDT, R., '17. Crossing over ohne Chiasmatypie? Genetics, 2: 82.

GONZALEZ, B. M., '23. Experimental studies on the duration of life. VIII. The influence upon duration of life of certain mutant genes of Drosophila melanogaster. Amer. Nat., 57: 289—325.

GOWEN, J. W., '19. A biometrical study of crossing over. On the mechanism of crossing over in the third chromosome of Drosophila melanogaster. Genetics, 4: 205—250.

GOWEN, M. S., and J. W. GOWEN, '22. Complete linkage in Drosophila melanogaster. Amer. Nat., 56: 286—288.

GRIMSHAW, P., '01. Diptera. Fauna Hawaiiensis, 3: 1—86 pl. I—III.

GUYÉNOT, E., '13. Études biologiques sur une mouche, Drosophila ampelophila Löw
- I. Possibilité de vie aseptique pour l'individu et la lignée. C. R. Soc. Biol., 74: 97.
- II. Rôle des levures dans l'alimentation. Ibid., 178.
- III. Changement de milieu et adaptation. Ibid., 223.
- IV. Nutrition des larves et fécondité. Ibid., 270.
- V. Nutrition des adultes et fécondité. Ibid., 332.
- VI. Resorption des spermatozoides et avortement des oeufs. Ibid., 389.
- VII. Le déterminisme de la ponte. Ibid., 443.

GUYÉNOT, E., '14. Action des rayons ultra-violets, sur Drosophila ampelophila Löw. Bull. Scient. France Belg., 48: 160—169.

GUYÉNOT, E., '17. Recherches sur la vie aseptique et le developpement d'un organisme en fonction du milieu. Thesis, Paris. 330 pp. + 4 plates.

GUYÉNOT, E., '20. Recherches sur un cas d'hérédité "sex-linked", la Drosophile a oeil "barred". Mém. Soc. Phys. Hist. Nat. Genève, 39, fasc. 5: 261—272.

HALDANE, J. B. S., '19. The combination of linkage values, and the calculation of distances between the loci of linked factors. Journ. Genet., 8: 299—309.

HERSH, A. H., '24. The effect of temperature upon the heterozygotes in the bar series of Drosophila. Journ. Exp. Zool., 39: 55—71.

HERSH, R. K., '24. The effect of temperature upon the full-eyed race of Drosophila. Jour. Exp. Zool., 39: 43—53.

HOGE, M. A., '15*a*. The influence of temperature on the development of a Mendelian character. Jour. Exp. Zool., 18: 241.

HOGE, M. A., '15*b*. Another gene in the fourth chromosome of Drosophila. Amer. Nat., 49: 47.

HOLMES, C. D., '10. The effect of starvation for five successive generations on the sex ratio in Drosophila ampelophila. Indiana Univ. Bull., 8: 16.

HOWARD, L. O., '00. A contribution to the study of the insect fauna of human excrement. Proc. Wash. Acad. Sci., 2: 541.

HUETTNER, A. F., '23. The origin of the germ cells in Drosophila melanogaster. Journ. Morphol., 37: 385—423.

HYDE, R. R., '13. Inheritance of the length of life in Drosophila ampelophila. Ind. Acad. Sci. Rept. '13: 113.

HYDE, R. R., '14. Fertility and sterility in Drosophila ampelophila.

I. Sterility in Drosophila with especial reference to a defect in the female and its behavior in heredity. Journ. Exp. Zool., 17: 141.

II. Fertility in Drosophila and its behavior in heredity. Journ. Exp. Zool., 17: 173.

III. Effects of crossing on fertility in Drosophila. Ibid., 343.

IV. Effects on fertility of crossing within and without an inconstant stock of Drosophila. Ibid., 356.

HYDE, R. R., '15*a*. The origin of a new eye color in Drosophila repleta, and its behavior in heredity. Amer. Nat., 49: 183.

HYDE, R. R., '15*b*. A wing mutation in a new species of Drosophila. Amer. Nat., 49:185.

HYDE, R. R., '16. Two members of a sex-linked multiple (sextuple) allelomorph system. Genetics, 1: 535.

HYDE, R. R., '21*a*. A high fecundity record for Drosophila melanogaster. Proc. Ind. Acad. Sci., '20: 259—260.

HYDE, R. R., 21*b*. Segregation and recombination of the genes for tinged, blood, buff, and coral in Drosophila melanogaster. Proc. Indiana Acad. Sci. '20: 291—300.

HYDE, R. R., '21*c*. Behavior of the gene for the mutant curved of Drosophila melanogaster in crosses involving genes in the same and other chromosomes. Proc. Indiana Acad. Sci. '20: 301—313.

HYDE, R. R., '22. An eyeless mutation in Drosophila hydei. Genetics, 7: 319—334.

HYDE, R. R., and H. M. POWELL, '16. Mosaics in Drosophila ampelophila. Genetics, 1: 581—583.

IVES, J. D., '21. Cross-over values in the fruit-fly, Drosophila ampelophila when the linked factors enter in different ways. Amer. Nat., 55: 571—573.

JENNINGS, H. S., '17*a*. Modifying factors and multiple allelomorphs in relation to the results of selection. Am. Nat., 51: 301—306.

JENNINGS, H. S., '17*b*. Observed changes in hereditary characters in relation to evolution. Journ. Wash. Acad. Sci., 7: 281—301.

JENNINGS, H. S., '18. Disproof of a certain type of theories of crossing over between chromosomes. Amer. Nat., 52: 247.

JEPSON, F. P., '17. Report, Division of Entomology. Dept. Agric. Fiji, Ann. Rept. for 1916. Suva ,pp. 16—25.

JOHNSON, C. W., '13. The distribution of some species of Drosophila. Psyche, 20: 202—204.

JUST, G., '20. Der Nachweis von Mendel-Zahlen bei Formen mit niedriger Nachkommenzahl. Eine empirische Prüfung der Geschwister- und Probandenmethode Weinbergs auf Grund Kreuzungsversuchen mit Drosophila ampelophila. Arch. Mikr. Anat., 94: 604—652.

KEILIN, D., '15. Recherches sur les larves de diptères cyclorhaphes. Bull. Sci. France Belg., 49: 15—198. 16 plates.

KNAB, F., '12. Drosophila repleta Wollaston. Psyche, 19: 106—108.

KRAFKA, Jr., J., '20*a*. The effect of temperature upon facet number in the bar-eyed mutant of Drosophila. Journ. Gen. Physiol., 2.
Part I, pp. 409—432.
Part II, pp. 433—444.
Part III, pp. 445—464.

KRAFKA, J., '20*b*. Environmental factors other than temperature affecting facet number in the bar-eyed mutant of Drosophila. Journ. Gen. Physiol., 3: 207—210.

KRÖBER, O., '12. Beitrag zur Biologie der Drosophilinae. Zeits. wiss. Insektenbiol., 8: 235—236. Berichtigung, p. 329.

LAMB, C. G., '14. Reports of the Percy Sladen Trust Expedition. Diptera. Trans. London Linn. Soc. 2 ser., 16: 307—372, pl. 19—21.

LAMB, C. G., '18. On a parasitic Drosophila from Trinidad. Bull. Ent. Research. 9: 157—162.

LANCEFIELD, D. E., '18*a*. Three mutations in previously known loci. Amer. Nat., 52: 264—269.

LANCEFIELD, D. E., '18*b*. An autosomal bristle modifier, affecting a sex-linked character. Amer. Nat., 52: 462—464.

LANCEFIELD, D. E., '18*c*. A case of abnormal inheritance in Drosophila melanogaster. Amer. Nat., 52: 556—558.

LANCEFIELD, D. E.,'18*d*. Scarlet, an autosomal eye color identical with sex-linked vermilion. Biol. Bull., 35: 207—210.

LANCEFIELD, D. E., '19. Two sex-linked lethals of simultaneous appearance in Drosophila obscura. Soc. Exp. Biol. Med., 17: 56—77.

LANCEFIELD, D. E., '22. Linkage relations of the sex-linked characters in Drosophila obscura. Genetics, 7: 335—384.

LANCEFIELD, R. C., and C. W. METZ, '21. Non-disjunction and the chromosome relationships of Drosophila willistoni. Proc. Nat. Acad. Sci., 7: 225—229.

LANCEFIELD, R. C., and C. W. METZ, '22. The sex-linked group of characters in Drosophila willistoni. Amer. Nat., 56: 211—241.

LIFF, J., '15. Data on a peculiar Mendelian ratio in Drosophila ampelophila. Amer. Nat., 49—97.

LIPPINCOTT, W. A., '18. The factors for yellow in mice and notch in Drosophila. Amer. Nat., 52: 364—365.

LITTLE, C. C., '21. Non-disjunction of the fourth chromosome of Drosophila. Science, n. s., 53: 167.

LITTLE, C. C., and E. E. JONES, '23. The effect of selection upon a Mendelian ratio. Genetics, 8: 1—26.

LOEB, J., '15*a*. The simplest constituents required for growth and the completion of the life cycle in an insect (Drosophila). Science, n. s., 41: 169—170.

LOEB, J., '15*b*. The salts required for the development of insects, Jour. Biol. Chem., 23: 431.

LOEB, J., and F. W. BANCROFT, '11. Some experiments in the production of mutants in Drosophila. Science, n. s., 33: 781—783.

LOEB, J., and J. H. NORTHROP, '16*a*. Nutrition and evolution. Jour. Biol. Chem., 27: 309—312.

LOEB, J., and J. H. NORTHROP, '16*b*. Is there a temperature coefficient for the duration of life? Proc. Nat. Acad. Sci., 2: 456—457.

LOEB, J., and J. H. NORTHROP, '17*a*. What determines the duration of life in metazoa? Proc. Nat. Acad. Sci., 3: 382—386.

LOEB, J., and J. H. NORTHROP, '17*b*. On the influence of food and temperature on the duration of life. Jour. Biol. Chem., 32: 103—121.

LOEW, H., 1862. Diptera americae septentrionalis indigena. II. Berl. ent. Zeit. (Centuria).

LUTZ, F. E., '11. Experiments with Drosophila ampelophila concerning evolution. Carnegie Inst. Wash. publ., 143.

LUTZ, F. E., '12. Inheritance of abnormal wing-venation in Drosophila. Proc. 7 Int. Zool. Congr. Boston, pp. 411—419.

LUTZ, F. E., '13. Experiments concerning the sexual difference in the wing length of Drosophila ampelophila. Jour. Exp. Zool., 14: 267—273.

LUTZ, F. E., '14. Biological notes concerning Drosophila ampelophila. Jour. N. Y. Ent. Soc., 22: 134—138.

LUTZ, F. E., '15. Experiments with Drosophila ampelophila concerning natural selection. Bull. Amer. Mus. Nat. Hist., 35: 605—624.

LUTZ, F. E., and F. K. Richtmeyer, '22. The reaction of Drosophila to ultraviolet. Science, 55: 519.

LYNCH, C. J., '19. An analysis of certain cases of intra-specific sterility. Genet., 4: 501—533.

MAC DOWELL, E. C., '15. Bristle inheritance in Drosophila, I. Extra bristles. Jour. Exp. Zool., 19: 61.

MAC DOWELL, E. C. '17*a*. Bristle inheritance in Drosophila. II. Selection. Journ. Exp. Zool., 23: 109—146.

MAC DOWELL, E. C. '17*b*. The bearing of selection experiments with Drosophila upon the frequency of germinal changes. Proc. Nat. Acad. Sci., 3: 291—297.

MAC DOWELL, E. C. '20. Bristle inheritance in Drosophila, III. Correlation. Journ. Exp. Zool., 30: 419—460.

MC EWEN, R. S., '17. The reactions to light and to gravity in Drosophila and its mutants. Journ. Exp. Zool., 25: 49—106.

MALLOCH, J. R., '21. Some notes on Drosophilidae. Entom. News, 32: 311—312.

MANN, M. C., '23*a*. The occurrence and hereditary behavior of two new dominant mutations in an inbred strain of Drosophila melanogaster. Genetics, 8: 27—36.

MANN, M. C., '23*b*. A demonstration of the stability of the genes of an inbred stock of Drosophila melanogaster under experimental conditions. Journ. Exp. Zool., 38: 213—244.

MARSHALL, W. W., and H. J. MULLER, '17. The effect of long-continued heterozygosis on a variable character in Drosophila. Journ. Exp. Zool., 22: 457—470.

MARTELLI, G., '10. Notize sulla Drosophila ampelophila Lw. Boll. Lab. zool. gen. agrar. Portici, 4: 163—174.

MAVOR, J. W., '21*a*. Elimination of the X-chromosome from the egg of Drosophila melanogaster by X-rays. Proc. Soc. Exp. Biol. Med., 18: 303—304.

MAVOR, J. W., '21*b*. On the elimination of the X-chromosome from the egg of Drosophila melanogaster by X-rays. Science, n. s., 54: 277—279.

MAVOR, J. W., '22. The production of non-disjunction by X-rays. Science, n. s., 55: 295—297.

MAVOR, J. W., '23*a*. X-rays and the sex chromosomes. Science, n. s., 57: 503—504.

MAVOR, J. W., '23*b*. An effect of X-rays on crossing-over in Drosophila. Proc. Soc. Exp. Biol. Med., 20: 335—338.

MAVOR, J. W., '23*c*. An effect of X-rays on the linkage of Mendelian characters in the first chromosome of Drosophila. Genetics, 8: 355—366.

MAVOR, J. W., '23*d*. Studies on the biological effects of X-rays. Amer. Jour. Roentgenology and Radium Ther., 10: 968—974.

MAVOR, J. W., '24. The production of non-disjunction by X-rays. Journ. Exper. Zool. 39: 381—432.

MAVOR, J. W., and H. K. SVENSON, '23. X-rays and crossingover. Science, n. s., 58: 124—126.

MAVOR, J. W., and H. K. SVENSON, '24. An effect of X-rays on the linkage of Mendelian characters in the second chromosome of Drosophila melanogaster. Genetics, 9: 70—89.

MAY, H. G., '17*a*. The appearance of reverse mutations in the bar-eyed race of Drosophila under experimental control. Proc. Nat. Acad. Sci., 3: 544—545.

MAY, H. G., '17*b*. Selection for higher and lower facet numbers in the bar-eyed race of Drosophila and the appearance of reverse mutations. Biol. Bull., 33: 361—395.

MAYER, P., 1879. Zur Lehre von den Sinnesorganen bei den Insecten. Zool. Anz., 2: 182—183.

MEIGEN, J. W., 1830. Systematische Beschreibung der bekannten europäischen zweiflügeligen Insekten. 6: 81.

DE MEIJERE, J. C. H., '11. Studien über südostasiatische Dipteren. VI. Tijd. v. Ent., 54: 258—432.

DE MEIJERE, J. C. H., '16. Beiträge zur Kenntnis der Dipteren-Larven und -Puppen. Zool. Jahrb. Abt. Syst., 40: 177—322.

METZ, C. W., '14*a*. An apterous Drosophila and its genetic behavior. Amer. Nat., 48: 675—711.

METZ, C. W., '14*b*. Chromosome studies in the Diptera. I. A preliminary survey of five different types of chromosome groups in the genus Drosophila. Journ. Exp. Zool., 17: 45—49.

METZ, C. W., '16*a*. Chromosome studies in the Diptera.
II. The paired association of chromosomes in the Diptera and its significance. Journ. Exp. Zool., 21: 213.
III. Additional types of chromosome groups in the Drosophilidae. Amer. Nat. 50: 587—599.

METZ, C. W., '16*b*. Mutations in three species of Drosophila. Genetics, 1: 591—607.

METZ, C. W., '16*c*. Linked Mendelian characters in a new species of Drosophila. Science, n. s., 44: 431—432.

METZ, C. W., '18. The linkage of eight sex-linked characters in Drosophila virilis. Genetics, 3: 107—134.

METZ, C. W., '20*a*. Correspondence between chromosome number and linkage groups in Drosophila virilis. Science, n. s., 51: 417—418.

METZ, C. W., '20*b*. The arrangement of genes in Drosophila virilis. Proc. Nat. Acad. Sci., 6: 164—166.

METZ, C. W., '20*c*. Observations on the sterility of mutant hybrids in Drosophila virilis. Proc. Nat. Acad. Sci., 6: 421—423.

METZ, C. W., '23. A note on the effects of temperature on the mutant characters "bent" in Drosophila virilis and Drosophila melanogaster. Proc.Soc.Exp. Biol. Med., 20: 305—310.

METZ, C. W., and C. B. BRIDGES, '17. Incompatibility of mutant races in Drosophila. Proc. Nat. Acad. Sci., 3: 673—678.

METZ, C. W., and R. M. FERRY, '23. The parallel characters "crossveinless" and "vermilion" in Drosophila willistoni. Amer. Nat., 57: 381—384.

METZ, C. W., and B. S. METZ, '15. Mutations in two species of Drosophila. Amer. Nat., 49: 187—189.

METZ, C. W., and M. S. MOSES, '23. Chromosomes of Drosophila. Chromosome relationships and genetic behavior in the genus Drosophila. I. A comparison of the chromosomes of different species of Drosophila. Journ. Hered., 14: 195—204.

METZ, C. W., M. S. MOSES, and E. D. MASON, '23. Genetic studies on Drosophila virilis with considerations on the genetics of other species of Drosophila. Carnegie Inst. Washington, publ. 328, 94 pp., 5 Pl.

MOENKHAUS, W. J., '11. The effects of inbreeding and selection on the fertility, vigor, and sex-ratio for Drosophila ampelophila. Jour. Morph., 22: 123.

MOHR, O. L., '19. Character changes caused by mutation of an entire region of a chromosome in Drosophila. Genetics, 4: 275—282.

MOHR, O. L., '21. Den Morgan'ske skole og dens betydning for den moderne arvelighedsforskning. Nord. Jordbrugsforsk. Foren. Kongr. København, '21: 234—277.

MOHR, O. L., '22*a*. Cases of mimic mutations and secondary mutations in the X-chromosome of Drosophila melanogaster. Zeits. ind. Abst. Vererb., 28: 1—22.

MOHR, O. L., '22*b*. Ö. Winge's paper on "The interaction between two closely linked lethals in Drosophila as the cause of the apparent constancy of the mutant "spread". Genetica, 4: 457—460.

MOHR, O. L., '23*a*. A somatic mutation in the singed locus of the X-chromosome in Drosophila melanogaster. Hereditas, 4: 142—160.

MOHR, O. L., '23*b*. Das Deficiency-Phänomen bei Drosophila melanogaster. Zeits. ind. Abst. Vererb., 30: 279—283.

MOHR, O. L., '23*c*. Modifications of the sex-ratio through a sex-linked semi-lethal in Drosophila melanogaster. (Besides notes on an autosomal section deficiency). Studia Mendeliana, Brünn, pp. 1—22.

MOHR, O. L., '23*d*. A genetic and cytological analysis of a section deficiency involving four units of the X-chromosome in Drosophila melanogaster. Zeits. ind. Abst. Vererb., 32: 108—232.

MOHR, O. L., and A. H. STURTEVANT, '19. A semi-lethal in Drosophila funebris that causes an excess of males. Proc. Soc. Exper. Biol. Med., 16: 95—96.

MORGAN, L. V., '22. Non-criss-cross inheritance in Drosophila melanogaster. Biol. Bull., 42: 267—274.

MORGAN, T. H., '10*a*. Hybridization in a mutating period in Drosophila. Proc. Soc. Exp. Biol. Med., 7: 160—161.

MORGAN, T. H., '10*b*. Sex limited inheritance in Drosophila. Science, n. s., 32: 120—122.

MORGAN, T. H., '10*c*. The method of inheritance of two sex-limited characters in the same animal. Proc. Soc. Exp. Biol. Med., 8: 17—19.

MORGAN, T. H., '11*a*. An alteration of the sex-ratio induced by hybridization. Proc. Soc. Exp. Biol. Med., 8: 82—83.

MORGAN, T. H., '11*b*. A dominant sex-limited character. Proc. Soc. Exp. Biol. Med., 9: 14—15.

MORGAN, T. H., '11*c*. The origin of nine wing-mutations in Drosophila. Science, n. s., 33: 496—499.

MORGAN, T. H., '11*d*. The application of the conception of pure lines to sex-limited inheritance and to sexual dimorphism. Amer. Nat., 45: 65—78.

MORGAN, T. H., '11*e*. Random segregation versus coupling in Mendelian inheritance. Science, n. s., 34: 384.

MORGAN, T. H., '11*f*. Chromosomes and associative inheritance. Science, n. s., 34: 636—638.

MORGAN, T. H., '11*g*. The origin of five mutations in eye-color in Drosophila and their modes of inheritance. Science, n. s., 33: 534—537.

MORGAN, T. H., '11*h*. An attempt to analyze the constitution of the chromosomes on the basis of sex-limited inheritance in Drosophila. Journ. Exp. Zool., 11: 365—413.

MORGAN, T. H., '12*a*. Further experiments with mutations in eye-color of Drosophila: the loss of the orange factor. Jour. Acad. Nat. Sci. Phila., 15: 321—346.

MORGAN, T. H., '12*b*. Eight factors that show sex-linked inheritance in Drosophila. Science, n. s., 35: 472—473.

MORGAN, T. H., '12*c*. The masking of a Mendelian result by the influence of the environment. Proc. Soc. Exp. Biol. Med., 9: 73—74.

MORGAN, T. H., '12*d*. Heredity of body color in Drosophila. Journ. Exp. Zool., 13: 27—43.

MORGAN, T. H., '12*e*. A modification of the sex-ratio, and of other ratios, in Drosophila through linkage. Zeits. Abst. Vererb., 7: 323—345.

MORGAN, T. H., '12*f*. The explanation of a new sex-ratio in Drosophila. Science, n. s., 36: 718—719.

MORGAN, T. H., '12*g*. Complete linkage in the second chromosome of the male. Science, n. s., 36: 719—720.

MORGAN, T. H., '13*a*. Heredity and sex. Columbia University Press.

MORGAN, T. H., '13*b*. Factors and unit characters in Mendelian heredity. Amer. Nat., 47: 5—16.

MORGAN, T. H., '13*c*. Simplicity versus adequacy in Mendelian formulae. Amer. Nat., 47: 372—374.

MORGAN, T. H., '14*a*. The mechanism of heredity as indicated by the inheritance of linked characters. Pop. Sci. Mo. '14: 5—16.

MORGAN, T. H., '14*b*. Mosaics and gynandromorphs in Drosophila. Proc. Soc. Exp. Biol. Med .,11: 171—172.

MORGAN, T. H., '14*c*. A third sex-linked lethal factor in Drosophila. Jour. Exp. Zool., 17: 315—324.

MORGAN, T. H., '14*d*. The failure of ether to produce mutations in Drosophila. Amer. Nat., 48: 705—711.

MORGAN, T. H., '14*e*. No crossing over in the male of Drosophila of genes in the second and third pairs of chromosomes. Biol. Bull., 26: 195—204.

MORGAN, T. H., '14*f*. Another case of multiple allelomorphs in Drosophila. Biol. Bull., 26: 231—244.

MORGAN, T. H., '14*g*. Two sex-linked lethal factors in Drosophila and their influence on the sex-ratio. Journ. Exp. Zool., 17: 81—122.

MORGAN, T. H., '15*a*. The infertility of rudimentary winged females of Drosophila ampelophila. Amer. Nat., 49: 240—250.

MORGAN, T. H., '15*b*. The constitution of the hereditary material. Proc. Amer. Phil. Soc., 54: 143—153.

MORGAN, T. H., '15*c*. The role of the environment in the realization of a sex-linked Mendelian character in Drosophila. Amer. Nat., 49: 385—429.

MORGAN, T. H., '15*d*. Localization of the hereditary material in the germ cells. Proc. Nat. Acad. Sci., 1: 420—429.

MORGAN, T. H., '16. A critique of the theory of evolution. Princeton Press.

MORGAN, T. H., '17*a*. An examination of the so-called process of contamination of the genes. Anat. Record, 11: 503.

MORGAN, T. H., '17*b*. The theory of the gene. Amer. Nat., 51: 513—544.

MORGAN, T. H., '18*a*. Concerning the mutation theory. Scient. Mo., 5: 385—405.

MORGAN, T. H., '18*b*. Changes in factors through selection. Scient. Mo., 5: 549—559.

MORGAN, T. H., '18*c*. Evolution by mutation. Scient. Mo., 6: 46—53.

MORGAN, T. H., '19*a*. A demonstration of genes modifying the character "notch". Carnegie Inst. Wash. publ. 278: 343—388.

MORGAN, T. H., '19*b*. The physical basis of heredity. Philadelphia. 305 pp.

MORGAN, T. H., '22*a*. The mechanism of heredity. Nature 109: 241—244, 275—278, 312—313.

MORGAN, T. H., '22*b*. On the mechanism of heredity. Croonian Lecture. Proc. Roy. Soc. B., 94: 162—197.

MORGAN, T. H., and C. B. BRIDGES, '13. Dilution effects and bicolorism in certain eye colors of Drosophila. Journ. Exp. Zool., 15: 429—466.

MORGAN, T. H., and C. B. BRIDGES, '16. Sex-linked inheritance in Drosophila. Carnegie Inst. Wash., publ. 237, 87 pp. + 2 plates.

MORGAN, T. H., and C. B. BRIDGES, '19*a*. The construction of chromosome maps. Proc. Soc. Exp. Biol. Med., 16: 96—97.

MORGAN, T. H., and C. B. BRIDGES, '19*b*. The origin of gynandromorphs. Carn. Inst. Wash., publ., 278: 1—122.

MORGAN, T. H., and C. B. BRIDGES, '19*c*. The inheritance of a fluctuating character. Journ. Gen. Physiol., 1: 639—643.

MORGAN, T. H., and E. CATTELL, '12. Data for the study of sex-linked inheritance in Drosophila. Journ. Exp. Zool., 13: 79—101.

MORGAN, T. H., and E. CATTELL, '13. Additional data for the study of sex-linked inheritance in Drosophila. Journ. Exp. Zool., 14: 33—42.

MORGAN, T. H., and C. J. LYNCH, '12. The linkage of two factors in Drosophila that are not sex-linked. Biol. Bull., 23: 174—182.

MORGAN, T. H., and H. H. PLOUGH, '15. The appearance of known mutations in other mutant stocks. Amer. Nat., 49: 318—319.

MORGAN, T. H., A. H. STURTEVANT, and C. B. BRIDGES, '20. The evidence for the linear order of the genes. Proc. Nat. Acad. Sci., 6: 162—164.

MORGAN, T. H., A. H. STURTEVANT, H. J. MULLER, and C. B. BRIDGES, '15. The mechanism of Mendelian heredity. Holt and Co. 262 pp.

MORGAN, T. H., and S. C. TICE, '14. The influence of the environment on the size of expected classes. Biol. Bull., 26: 213—220.

MULLER, H. J., '14*a*. A factor for the fourth chromosome of Drosophila. Science, n. s., 39: 906.

MULLER, H. J., '14*b*. A gene for the fourth chromosome of Drosophila. Journ. Exp. Zool., 17: 325.

MULLER, H. J., '16. The mechanism of crossing over. Amer. Nat., 50: 193—221 284—305, 350—366, 421—434.

MULLER, H. J., '17. An Oenothera-like case in Drosophila. Proc. Nat. Acad. Sci., 3: 619—626.

MULLER, H. J., '18. Genetic variability, twin hybrids and constant hybrids, in a case of balanced lethal factors. Genetics, 3: 422—499.

MULLER, H. J., '20*a*. Are the factors of heredity arranged in a line? Amer. Nat., 54: 97—121.

MULLER, H. J., '20*b*. Further changes in the white-eye series of Drosophila and their bearing on the manner of occurrence of mutation. Journ. Exp. Zool., 31: 443—473.

MULLER, H. J., '23. Recurrent mutations of normal genes of Drosophila not caused by crossing over. Anat. Rec., 26: 397—398.

MULLER, H. J., and E. ALTENBURG, '19. The rate of change of hereditary factors in Drosophila. Proc. Soc. Exp. Biol. Med., 17: 10—14.

NACHTSHEIM, H., '19*a*. Die Analyse der Erbfaktoren bei Drosophila und deren zytologische Grundlage. Zeitschr. ind. Abst. Vererb., 20: 118—156.

NACHTSHEIM, H., '19*b*. Der Mechanismus der Vererbung. Naturwiss. Wochenschr., n. f., 18: 105—114.

NONIDEZ, J. F., '20. The internal phenomena of reproduction in Drosophila. Biol. Bull., 39: 207—230.

NORTHROP, J. H., '17*a*. The effect of prolongation of the period of growth on the total duration of life. Jour. Biol. Chem., 32: 123—126.

NORTHROP, J. H., '17*b*. The rôle of yeast in the nutrition of an insect (Drosophila). Journ. Biol. Chem., 30: 181—187.

NORTHROP, J. H., '20. Concerning the hereditary adaptation of organisms to higher temperature. Journ. Gen. Physiol., 2: 313—318.

OLDENBERG, L., '14. Beitrag zur Kenntnis der europäischen Drosophiliden. Arch. Naturgesch., 80, A; 2: 1.

PAYNE, F., '10. Forty-nine generations in the dark. Biol. Bull., 18: 188—190.

PAYNE, F., '11. Drosophila ampelophila bred in the dark for sixty-nine generations. Biol. Bull., 21: 297—301.

PAYNE, F., '18*a*. The effect of artificial selection on bristle number in Drosophila ampelophila and its interpretation. Proc. Nat. Acad. Sci., 4: 55.

PAYNE, F., '18*b*. An experiment to test the nature of the variations on which selection acts. Indiana Univ. Studies, 5: 1—45.

PAYNE, F., '20. Selection for high and low bristle number in the mutant strain "reduced". Genetics, 6: 501—542.

PAYNE, F., '21. The heredity of orange eye color in Drosophila melanogaster. Amer. Nat., 55: 377—381.

PAYNE, F., and M. DENNY, '22. A gynandromorph in Drosophila melanogaster. Amer. Nat., 46: 383—384.

PEARL, R., '22. Experimental studies on the duration of life. VI. A comparison of the laws of mortality in Drosophila and in man. Amer. Nat., 56: 398—405.

PEARL, R., and S. PARKER, '21. Experimental studies on the duration of life. I. Introductory discussion of the duration of life in Drosophila. Amer. Nat., 55: 481—509.

PEARL, R., and S. L. PARKER, '22. Experimental studies on the duration of life. II. Hereditary differences in duration of life in line-bred strains of Drosophila. Amer. Nat., 56: 174—187.

III. The effect of successive etherizations on the duration of life of Drosophila. Amer. Nat., 56: 273—280.

IV. Data on the influence of density of population on duration of life in Drosophila. Amer. Nat., 56: 312—321.

V. On the influence of certain environmental factors on duration of life in Drosophila. Amer. Nat., 56: 385—398.

PEARL, R., and S. L. PARKER '22. On the influence of density of population upon the rate of reproduction in Drosophila. Proc. Nat. Acad. Sci., 8: 212—219.

PEARL, R., and S. L. PARKER '22. New experimental data on the influence of density of population upon duration of life in Drosophila. Amer. Jour. Hygiene, 3: 94—97.

PEARL, R., and S. L. PARKER '24. Experimental studies on the duration of life. IX. New life tables for Drosophila. Amer. Nat., 58: 71—82. X. The duration of life of Drosophila melanogaster in the complete absence of food. Amer. Nat. 58: 193—218.

PEARL, R., S. L. PARKER, and B. M. GONZALEZ, '23. Experimental studies on the duration of life. VII. The Mendelian inheritance of duration of life in crosses of wild type and quintuple stocks of Drosophila melanogaster. Amer. Nat., 57: 153—192.

PERKINS, R. C. L., '13. Drosophilidae, in Fauna Hawaiiensis, Introduction. 1: 189.

PLOUGH, H. H., '17*a*. The effect of temperature on linkage in the second chromosome of Drosophila. Proc. Nat. Acad. Sci., 3: 553—555.

PLOUGH, H. H., '17*b*. The effect of temperature on crossingover in Drosophila. Journ. Exp. Zool., 24: 147—209.

PLOUGH, H. H., '19. Linear arrangement of genes and double crossing over. Proc. Nat. Acad. Sci., 5: 167—168.

PLOUGH, H. H., '21. Further studies on the effect of temperature on crossing over. Journ. Exp. Zool., 32: 187—202.

PLOUGH, H. H., '24. Radium radiations and crossing over. Amer. Nat., 58: 85—87.

PLOUGH, H. H., and M. B. STRAUSS, '23. Experiments on toleration of temperature by Drosophila. Journ. Gen. Physiol., 6: 167—176.

QUACKENBUSH, L. S., '10. Unisexual broods of Drosophila. Science, n. s., 32: 183—185.

RAWLS, E., '13. Sex ratios in Drosophila ampelophila. Biol. Bull., 24: 115.

REEVES, EDNA M., '16. The inheritance of extra bristles in Drosophila melanogaster Meig. Univ. Calif. Publ. Zool., 13: 495.

RICHARDS, M. H., '18. Two new eye colors in the third chromosome of Drosophila melanogaster. Biol. Bull., 35: 199—206.

RICHARDS, M. H., and E. Y. FURROW, '22. A preliminary report on the optic tract of eyeless flies. Univ. of Okla. Bull., 2: 41—45.

ROBERTS, E., '18. Fluctuations in a recessive Mendelian character and selection. Journ. Exp. Zool., 27: 157—192.

ROUBAUD, E., '10. Sur un nouveau Flagellé à forme trypanosome des Drosophiles d'Afrique, Cercoplasma drosophilae. C. R. Soc. Biol., 69: 554—556.

SAFIR, S. R., '13. A new eye-color mutation in Drosophila. Biol. Bull., 25: 45.

SAFIR, S. R., '16. Buff, a new allelomorph of white eye color in Drosophila. Genetics, 1: 584—590.

SAFIR, S. R., '20. Genetic and cytological examination of the phenomena of primary non-disjunction in Drosophila melanogaster. Genetics, 5: 459—487.

SCHINER, J. R., 1864. Fauna Austriaca. Diptera. Vienna. vol. 2.

SCHULZE, P., '12. Entwicklung von Drosophila rubrostriata Becker in Formol; ein Beitrag zur Kenntnis der Lebensweise der Drosophilalarven. Zool. Anz., 39: 199.

SEYSTER, E. W., '19. Eye facet number as influenced by temperature in the bar-eyed mutant of Drosophila melanogaster (ampelophila). Biol. Bull., 37: 168—182.

SILVESTRI, F., '17. Sulla Lonchaea aristella, damnosa alle inflorescenze e fruttescenze del caprifico e del fico. Boll. Lab. Zool. Gen. Agrar. Portici, 12:123—146.

STARK, M. B., '15. The occurrence of lethal factors in inbred and wild stocks of Drosophila. Journ. Exp. Zool., 19: 531—558.

STARK, M. B., '18. An hereditary tumor in the fruit fly, Drosophila. Journ. Cancer Research, 3: 279—300.

STARK, M. B., '19*a*. A benign tumor that is hereditary in Drosophila. Proc. Nat. Acad. Sci., 5: 573—580.

STARK, M. B., '19*b*. A benign tumor hereditary in Drosophila. Proc. Soc. Exper. Biol. Med., 17: 51—52.

STARK, M. B., '19*c*. An hereditary tumor. Journ. Exp. Zool., 27: 509—529.

STEVENS, N. M., '08. A study of the germ-cells of certain Diptera. Journ. Exp. Zool., 5: 359—374.

STEVENS, N. M., '12. The chromosomes in Drosophila ampelophila. Proc. 7 Int. Zool. Congr. Boston, pp. 380—381.

STRONG, L. C., '20. Roughoid, a mutant located to the left of sepia in the third chromosome of Drosophila melanogaster. Biol. Bull. 38: 33—37.

STURTEVANT, A. H., '13*a*. A third group of linked genes in Drosophila ampelophila. Science, n. s., 37: 990.

STURTEVANT, A. H., '13*b*. The linear arrangement of six sex-linked factors in Drosophila, as shown by their mode of association. Journ. Exp. Zool., 14: 43—59.

STURTEVANT, A. H., '13*c*. The Himalayan rabbit case, with some considerations on multiple allelomorphs. Amer. Nat., 47: 234—239.

STURTEVANT, A. H., '14. The reduplication hypothesis as applied to Drosophila. Amer. Nat., 48: 535—549.

STURTEVANT, A. H., '15*a*. The behavior of the chromosomes as studied through linkage. Zeits. Abst. Vererb., 13: 234—287.

STURTEVANT, A. H., '15*b*. Experiments on sex recognition and the problem of sexual selection in Drosophila. Jour. An. Behav., 5: 351—366.

STURTEVANT, A. H., '15*c*. A sex-linked character in Drosophila repleta. Amer. Nat., 49: 190.

STURTEVANT, A. H., '16. Notes on North American Drosophilidae with descriptions of twenty-three new species. Ann. Ent. Soc. Amer., 9: 323.

STURTEVANT, A. H., '17*a*. Crossing over without chiasmatype? Genetics, 2: 301—304.

STURTEVANT, A. H., '17*b*. An analysis of the effect of selection on bristle number in a mutant race of Drosophila. Anat. Record, 11: 504.

STURTEVANT, A. H., '17*c*. Genetic factors affecting the strength of linkage in Drosophila. Proc. Nat. Acad. Sci., 3: 555—558.

STURTEVANT, A. H., '18*a*. An analysis of the effects of selection. Carnegie Inst. Publ., 264 pp. 68 + 1 plate.

STURTEVANT, A. H., '18*b*. Acalypterae (Diptera) collected in Mobile County, Alabama. Jour. N. Y. Ent. Soc., 26: 34—40.

STURTEVANT, A. H., '18*c*. A synopsis of the Neartic species of the genus Drosophila (*Sensu lato*). Bull. Amer. Mus Nat. Hist., 38: 441—446.

STURTEVANT, A. H., '18*d*. A parallel mutation in Drosophila funebris. Science, 48: 72—73.

STURTEVANT, A. H., '18*e*. Flies of the genus Drosophila as possible disease carriers. Jour. Parasitol., 5: 84—85.

STURTEVANT, A. H., '19*a*. Inherited linkage variations in the second chromosome. Carnegie Inst. Wash. publ. 278: 305—341.

STURTEVANT, A. H., '19*b*. A new species closely resembling Drosophila melanogaster. Psyche 26: 153—155.

STURTEVANT, A. H., '20*a*. Intersexes in Drosophila simulans. Science, n. s., 51: 325—327.

STURTEVANT, A. H., '20*b*. The vermilion gene and gynandromorphism. Proc. Soc. Exp. Biol. Med., 17: 70—71.

STURTEVANT, A. H., '20*c*. The dipterous genus Zygothrica of Wiedemann. Proc. U. S. Nat. Mus., 58: 155—158.

STURTEVANT, A. H., '20*d*. Genetic studies on Drosophila simulans.

I. Introduction. Hybrids with D. melanogaster. Genetics, 5: 488—500.

STURTEVANT, A. H., '21*a*. Genetic studies on Drosophila simulans.

II. Sex-linked group of genes. Genetics, 6: 43—64.

III. Autosomal genes. General discussion. Genetics, 6: 179—207.

STURTEVANT, A. H., '21*b*. The North American species of Drosophila. Carnegie Inst. Wash., publ., 301. 150 pp., 3 pl.

STURTEVANT, A. H., '21*c*. Linkage variation and chromosome maps. Proc. Nat. Acad. Sci., 7: 181—183.

STURTEVANT, A. H., '21*d*. A case of rearrangement of genes in Drosophila. Proc. Nat. Acad. Sci., 7: 235—237.

STURTEVANT, A. H., C. B. BRIDGES, and T. H. MORGAN. '19. The spatial relations of genes. Proc. Nat. Acad. Sci., 5: 168—173.

STURTEVANT, A. H., and T. H. MORGAN, '23. Reverse mutation of the bar gene correlated with crossing over. Science, n. s., 57: 746—747.

THOMPSON, D. H., '20. A new type of sex linked lethal in Drosophila. Anat. Rec. 20: 215.

TICE, S. C., '14. A new sex-linked character in Drosophila. Biol. Bull., 26: 221.

UNWIN, E. E., '07. The vinegar fly (Drosophila funebris). Trans. Ent. Soc. London, p. 285—302.

VILLENEUVE, J., '13. Notes synonymiques. Wien. ent. Zeit., 32: 128.

WAALER, H. G. M., '21. The location of a new second chromosome eye colour gene in Drosophila melanogaster. Hereditas, 2: 391—394.

WARD, L., '23. The genetics of curly wing in Drosophila, another case of balanced lethal factors. Genetics, 8: 276—300.

WARREN, D. C., '17. Mutations in Drosophila busckii Coq. Amer. Nat., 51: 699—703.

WARREN, D. C., '18. The effect of selection upon the sex-ratio in Drosophila ampelophila. Biol. Bull., 34: 351—371.

WARREN, D. C., '20. Spotting inheritance in Drosophila busckii. Genetics, 5: 60—110.

WARREN, D. C., '24. Inheritance of egg size in Drosophila melanogaster. Genetics, 9: 41—69.

WEINSTEIN, A., '18. Coincidence of crossing over in Drosophila melanogaster (ampelophila). Genetics 3: 135—173.

WEINSTEIN, A., '20. Homologous genes and linear linkage in Drosophila virilis. Proc. Nat. Acad. Sci., 6: 626—639.

WEINSTEIN, A., '22. Crossing over, non-disjunction, and mutation in Drosophila virilis. Sigma Xi Quart., 10: 45—53.

WENTWORTH, E. N., '14. The segregation of fecundity factors in Drosophila. Jour. Genet., 3: 113—120.

WHITING, P. W., '13. Viability and coupling in Drosophila. Amer. Nat., 47: 508.

WILLISTON, S. W., '82. Drosophila ampelophila Loew. Canad. Entomol., 14: 138.

WILLISTON, S. W., '96. On the Diptera of St. Vincent. Trans. Entom. Soc. London, '96: 253—446.

WILSON, E. B., '13. Heredity and microscopical research. Science, n. s., 37: 814—826.

WILSON, E. B., '14. Croonian lecture: the bearing of cytological research on heredity. Proc. Roy. Soc. B, 88: 333—352.

WILSON, E. B. and T. H. MORGAN, '20. Chiasmatype and crossing-over. Am. Nat. 54: 193—219.

WINGE, Ö., '22. The interaction between two closely linked lethals in Drosophila as the cause of the apparent constancy of the mutant "spread". Genetica, 4: 321—338.

ZELENY, C., '18. Germinal changes in the bar-eyed race of Drosophila during the course of selection for facet number. Proc. Indiana Acad. Sci., '17: 73—77.

ZELENY, C., '19. A change in the bar gene of Drosophila involving further decrease in facet number and increase in dominance. Jour. Gen. Physiol., 2: 69—71.

ZELENY, C., '20. A change in the bar gene of Drosophila melanogaster involving further decrease in facet number and increase in dominance. Journ. Exp. Zool., 30: 293—324.

ZELENY, C., '21*a*. The direction and frequency of mutation in the bar-eye series of multiple allelomorphs of Drosophila. Journ. Exp. Zool., 34: 203—233.

ZELENY, C., '21*b*. Decrease in sexual dimorphism of bar-eye Drosophila during the course of selection for low and high facet number. Amer. Nat., 55: 404—411.

ZELENY, C., '22. The effect of selection for eye facet number in the white bar-eye race of Drosophila melanogaster. Genetics, 7: 1—115.

ZELENY, C., '23. The temperature coefficient of a heterozygote with an expression for the value of a germinal difference in terms of an environmental one. Biol. Bull., 44: 105—112.

ZELENY, C., and E. W. MATTOON, '15. The effect of selection upon the "bar-eye" mutant of Drosophila. Journ. Exp. Zool., 19: 515—529.

INDEX.